Seed Moisture

WITHDRAWN

Seed Moisture

Editors

Phillip C. Stanwood and Miller B. McDonald

Proceedings of a symposium sponsored by Divisions C-4 and C-2 of the Crop Science Society of America in Atlanta, Georgia, 30 Nov. 1987.

Organizing Committee
Phillip C. Stanwood
Miller B. McDonald

Editorial Committee
Phillip C. Stanwood
Miller B. McDonald
Eric E. Roos
Dennis M. TeKrony
S. H. West

Editor-in-Chief CSSA
C.W. Stuber

Managing Editor
S.H. Mickelson

CSSA Special Publication Number 14

Crop Science Society of America
Madison, Wisconsin, USA
1989

Cover Design:

The diagram on the front cover depicts representative physical, biochemical, and physiological changes associated with desiccation. The three vertical scales on the left present the physical parameters of desiccation. The first is water potential, defined as the chemical potential of water divided by the molar volume at atmospheric pressure; it is given in Pascals. The next scale is osmolarity; the osmolarities of blood, sea water and the water in the Great Salt Lake serve as reference points. The third scale is relative humidity (RH), or water activity (A_w); the values are entered on a log log-scale.

In the middle and right-hand sections of the diagram, some biological and physical changes are entered at the appropriate values on the dryness scales. The biological changes shown include wilting of plants, death of recalcitrant pollen and recalcitrant seeds, and the termination of metabolism in orthodox seeds. The values at which some representative microorganisms stop growing are depicted, as are the values at which nematodes become coiled and *Artemia* metabolism ceases.

The physical changes illustrated include the decrease in distance between membrane layers, to the point at which membranes collapse or form a Hex_{II} configuration, the folding of DNA molecules into liquid crystals, the freezing of membrane lipids, the cessation of the enzymic activity of lysozymes, and the point below which the dielectric reaches a minimum in wheat seeds. The changes in water potential of three types of soils during drying are also shown.

This diagram was designed by Joe Wolfe and A. Carl Leopold and reproduced with permission from Comstock Publishing Associated, Ithaca, NY (Leopold, 1986).

Crop Science Society of America, Inc.
677 South Segoe Road, Madison, WI 53711, USA

Copyright © 1989 by the Crop Science Society of America, Inc.

ALL RIGHTS RESERVED UNDER THE U.S. COPYRIGHT LAW
OF 1978 (P.L. 94-553)

Any and all uses beyond the limitations of the "fair use" provision of the law require written permission from the publisher(s) and/or the author(s); not applicable to contributions prepared by officers or employees of the U.S. Government as part of their official duties.

Library of Congress Cataloging-in-Publication Data

Seed moisture.

(CSSA special publication; no. 14)
Includes bibliographies.
1. Seeds—Moisture—Congresses. I. Stanwood, Phillip C. II. McDonald, M.B. III. Crop Science Society of America. Division C-4. IV. Crop Science Society of America. Division C-2. V. Series.
QK661.S418 1989 631.5'21   88-36212
ISBN 0-89118-525-9

Printed in the United States of America.

CONTENTS

Foreword vii
Preface ix
Contributors xi
Conversion Factors for SI and non-SI units xiii

1 Effects of Water on the Stability of Phospholipid Bilayers: The Problem of Imbibition Damage in Dry Organisms
   John H. Crowe, Lois M. Crowe, Folkert A. Hoekstra, and Christina Aurell Wistrom 1

2 Seed Moisture: Recalcitrant vs. Orthodox Seeds
   H. F. Chin, B. Krishnapillay, and P.C. Stanwood 15

3 Regulatory Roles for Desiccation and Abscisic Acid in Seed Development: A Comparison of the Evidence from Whole Seeds and Isolated Embryos
   Allison R. Kermode, Maurice Y. Oishi, and J. Derek Bewley 23

4 Moisture as a Regulator of Physiological Reaction in Seeds
   A. Carl Leopold and Christina W. Vertucci 51

5 Measurement of Seed Moisture
   D. F. Grabe 69

6 The Kinetics of Seed Imbibition: Controlling Factors and Relevance to Seedling Vigor
   Christina W. Vertucci 93

FOREWORD

The seed has long been revered as a storehouse of programmed plant development. Similarly, the seed is recognized as a living unit that respires and has the ability to exchange water with its environment. Seed of crop and non-crop species vary widely, and many can survive a wide range of environmental stress while retaining viability. This unique capability adds to the economic value of seed. For agricultural crops, where the seed is also used as a food or feed, the storage capacity includes a food quality component that is also sensitive to moisture.

It is clear that moisture content influences the capability of seed to survive and be vigorous in growth. Only recently, however, have the underlying physical, chemical, and physiological mechanisms been studied in detail in order to understand biological water in the seed and the water-embryo relationship. The new knowledge about these detailed and complex processes will give better insight to scientists, helping them to understand the pivotal points, those that regulate the process of moisture exchange and are subject to genetic or environmental manipulation. *Seed Moisture* contributes to that understanding and will serve as an important reference for researchers and educators.

Each of the authors has contributed valuable insight and a synthesis of a special topic. Phillip C. Stanwood and Miller B. McDonald deserve special recognition for organizing the symposium on seed moisture held at the 1987 annual meetings of the American Society of Agronomy and the Crop Science Society of America, and for carefully editing the manuscripts that ensued. The Society acknowledges the services of the capable authors, editors, and reviewers who contributed to the professionalism of this addition to the Special Publication series.

C. Jerry Nelson, *President,* 1988
Crop Science Society of America

PREFACE

One of the authors (H.F. Chin, Chapter 2) of this publication stated that "Water is life." While this statement appears mundane, when considered in the context of our lives and environment, it has profound and deep implications in everything we do. The topic of water and its role in physiological events has been the subject of several publications. For example, in a related publication on dry organisms (Leopold, 1986), the cover diagram considered measures of dryness, biological, and physical responses to water. This diagram emphasizes that many biological organisms, including seeds, can withstand extremely low levels of desiccation. Yet upon rehydration, these organisms regain full sensitivity to their environment, grow and complete their life cycle. This is an extraordinary event, but very common and from a technical point of view, poorly understood.

Seed moisture content plays a critical role in all aspects of seed science, technology, and commercial trade. In this publication, we explore the interesting facets of biological water in general and seed moisture content in particular. Underlying physiological aspects are addressed in terms of membrane systems (Chapter 1), physiological reaction in seeds (Chapter 4), and seed imbibition (Chapter 6). Differences between "orthodox" and "recalcitrant" seeds are surveyed in relation to moisture content (Chapter 2). The role of seed moisture content and development of the seed on the mother plant is reviewed in Chapter 3. Finally, methods and problems of measuring seed moisture content are explored in Chapter 5.

The objective of this publication was to survey what is known about seed moisture and hopefully complement interest and activity in this important area of seed technology. All facets could not be covered, but it is hoped that the topics presented here provide support for avenues already taken and furnish sign posts for new, innovative research and technical application.

A technical note: Some authors report moisture content on a fresh weight basis (fw) and some on a dry weight (dw) basis. The reader should be alerted to this, especially when making comparisons between studies. In all cases, we have attempted to keep the reporting units or ratios the same. Consequently, moisture content is reported on a range of 0 to 1.0. For example, 0.10 g H_2O g^{-1} dw would be 10% seed moisture content on a dry weight basis.

This special publication, *Seed Moisture,* is a product of a symposium presented at the 1987 Annual Meetings of the American Society of Agronomy in Atlanta, GA. The symposium was sponsored jointly by Div. C-4 (Seed Physiology, Production, and Technology) and Div. C-2 (Crop Physiology and Metabolism). The organizers would like to acknowledge the support of Northrup King Co., and Pioneer Hi-Bred International, Inc. We also acknowledge the cooperation of the authors, reviewers, and the support provided by the members of the respective divisions which made this publication possible.

Reference

Leopold, A.C. 1986. Membranes, metabolism and dry organisms. Comstock Publishing Associated, Cornell University Press, Ithaca, NY.

Phillip C. Stanwood and Miller B. McDonald
Organizers and Editors

CONTRIBUTORS

J. Derek Bewley	Professor, Department of Botany, University of Guelph, Guelph, Ontario, Canada
H. F. Chin	Professor, Department of Agronomy and Horticulture, Universiti Pertanian Malaysia, Serdang, Selangor, Malaysia
John H. Crowe	Professor, Department of Zoology, University of California, Davis, California
Lois M. Crowe	Associate Professor, Department of Zoology, University of California, Davis, California
D. F. Grabe	Professor of Agronomy, Crop Science Department, Oregon State University, Corvallis, Oregon
Folkert A. Hoekstra	Associate Professor of Plant Physiology, Department of Plant Physiology, Arboretumlaan, Wageningen, The Netherlands
Allison R. Kermode	Formerly Ph.D. Student, Department of Botany, University of Guelph, Guelph, Ontario, Canada. Currently Research Scientist, CSIRO Division of Plant Industry, Canberra, Australia
B. Krishnapillay	Research Fellow, Department of Agronomy and Horticulture, Universiti Pertanian Malaysia, Serdang, Selangor, Malaysia
A. Carl Leopold	W.C. Crocker Scientist, Boyce Thompson Institute, Cornell University, Ithaca, New York
Maurice Y. Oishi	Graduate Student, Department of Botany, University of Guelph, Guelph, Ontario, Canada
Phillip C. Stanwood	Research Agronomist, USDA-ARS, National Seed Storage Laboratory, Colorado State University, Fort Collins, Colorado
Christina W. Vertucci	Plant Physiologist, USDA-ARS, National Seed Storage Laboratory, Colorado State University, Fort Collins, Colorado
Christina Aurell Wistrom	Graduate Student, Department of Zoology, University of California, Davis, California

Conversion Factors for SI and non-SI Units

Conversion Factors for SI and non-SI Units

To convert Column 1 into Column 2, multiply by	Column 1 SI Unit	Column 2 non-SI Unit	To convert Column 2 into Column 1, multiply by
		Length	
0.621	kilometer, km (10^3 m)	mile, mi	1.609
1.094	meter, m	yard, yd	0.914
3.28	meter, m	foot, ft	0.304
1.0	micrometer, μm (10^{-6} m)	micron, μ	1.0
3.94×10^{-2}	millimeter, mm (10^{-3} m)	inch, in	25.4
10	nanometer, nm (10^{-9} m)	Angstrom, Å	0.1
		Area	
2.47	hectare, ha	acre	0.405
247	square kilometer, km^2 (10^3 m)^2	acre	4.05×10^{-3}
0.386	square kilometer, km^2 (10^3 m)^2	square mile, mi^2	2.590
2.47×10^{-4}	square meter, m^2	acre	4.05×10^3
10.76	square meter, m^2	square foot, ft^2	9.29×10^{-2}
1.55×10^{-3}	square millimeter, mm^2 (10^{-6} m)^2	square inch, in^2	645
		Volume	
9.73×10^{-3}	cubic meter, m^3	acre-inch	102.8
35.3	cubic meter, m^3	cubic foot, ft^3	2.83×10^{-2}
6.10×10^4	cubic meter, m^3	cubic inch, in^3	1.64×10^{-5}
2.84×10^{-2}	liter, L (10^{-3} m^3)	bushel, bu	35.24
1.057	liter, L (10^{-3} m^3)	quart (liquid), qt	0.946
3.53×10^{-2}	liter, L (10^{-3} m^3)	cubic foot, ft^3	28.3
0.265	liter, L (10^{-3} m^3)	gallon	3.78
33.78	liter, L (10^{-3} m^3)	ounce (fluid), oz	2.96×10^{-2}
2.11	liter, L (10^{-3} m^3)	pint (fluid), pt	0.473

CONVERSION FACTORS FOR SI AND NON-SI UNITS

Mass

To convert Column 1 into Column 2, multiply by	Column 1 SI Unit	Column 2 non-SI Unit	To convert Column 2 into Column 1, multiply by
2.20×10^{-3}	gram, g (10^{-3} kg)	pound, lb	454
3.52×10^{-2}	gram, g (10^{-3} kg)	ounce (avdp), oz	28.4
2.205	kilogram, kg	pound, lb	0.454
0.01	kilogram, kg	quintal (metric), q	100
1.10×10^{-3}	kilogram, kg	ton (2000 lb), ton	907
1.102	megagram, Mg (tonne)	ton (U.S.), ton	0.907
1.102	tonne, t	ton (U.S.), ton	0.907

Yield and Rate

0.893	kilogram per hectare, kg ha^{-1}	pound per acre, lb acre^{-1}	1.12
7.77×10^{-2}	kilogram per cubic meter, kg m^{-3}	pound per bushel, bu^{-1}	12.87
1.49×10^{-2}	kilogram per hectare, kg ha^{-1}	bushel per acre, 60 lb	67.19
1.59×10^{-2}	kilogram per hectare, kg ha^{-1}	bushel per acre, 56 lb	62.71
1.86×10^{-2}	kilogram per hectare, kg ha^{-1}	bushel per acre, 48 lb	53.75
0.107	liter per hectare, L ha^{-1}	gallon per acre	9.35
893	tonnes per hectare, t ha^{-1}	pound per acre, lb acre^{-1}	1.12×10^{-3}
893	megagram per hectare, Mg ha^{-1}	pound per acre, lb acre^{-1}	1.12×10^{-3}
0.446	megagram per hectare, Mg ha^{-1}	ton (2000 lb) per acre, ton acre^{-1}	2.24
2.24	meter per second, m s^{-1}	mile per hour	0.447

Specific Surface

10	square meter per kilogram, m^2 kg^{-1}	square centimeter per gram, cm^2 g^{-1}	0.1
1000	square meter per kilogram, m^2 kg^{-1}	square millimeter per gram, mm^2 g^{-1}	0.001

Pressure

9.90	megapascal, MPa (10^6 Pa)	atmosphere	0.101
10	megapascal, MPa (10^6 Pa)	bar	0.1
1.00	megagram per cubic meter, Mg m^{-3}	gram per cubic centimeter, g cm^{-3}	1.00
2.09×10^{-2}	pascal, Pa	pound per square foot, lb ft^{-2}	47.9
1.45×10^{-4}	pascal, Pa	pound per square inch, lb in^{-2}	6.90×10^3

continued on next page

Conversion Factors for SI and non-SI Units

To convert Column 1 into Column 2, multiply by	Column 1 SI Unit	Column 2 non-SI Unit	To convert Column 2 into Column 1, multiply by
	Temperature		
1.00 (K − 273)	Kelvin, K	Celsius, °C	1.00 (°C + 273)
(9/5 °C) + 32	Celsius, °C	Fahrenheit, °F	5/9 (°F − 32)
	Energy, Work, Quantity of Heat		
9.52×10^{-4}	joule, J	British thermal unit, Btu	1.05×10^{3}
0.239	joule, J	calorie, cal	4.19
10^{7}	joule, J	erg	10^{-7}
0.735	joule, J	foot-pound	1.36
2.387×10^{-5}	joule per square meter, J m^{-2}	calorie per square centimeter (langley)	4.19×10^{4}
10^{5}	newton, N	dyne	10^{-5}
1.43×10^{-3}	watt per square meter, W m^{-2}	calorie per square centimeter minute (irradiance), cal cm^{-2} min^{-1}	698
	Transpiration and Photosynthesis		
3.60×10^{-2}	milligram per square meter second, mg m^{-2} s^{-1}	gram per square decimeter hour, g dm^{-2} h^{-1}	27.8
5.56×10^{-3}	milligram (H$_2$O) per square meter second, mg m^{-2} s^{-1}	micromole (H$_2$O) per square centimeter second, μmol cm^{-2} s^{-1}	180
10^{-4}	milligram per square meter second, mg m^{-2} s^{-1}	milligram per square centimeter second, mg cm^{-2} s^{-1}	10^{4}
35.97	milligram per square meter second, mg m^{-2} s^{-1}	milligram per square decimeter hour, mg dm^{-2} h^{-1}	2.78×10^{-2}
	Plane Angle		
57.3	radian, rad	degrees (angle), °	1.75×10^{-2}

CONVERSION FACTORS FOR SI AND NON-SI UNITS

Electrical Conductivity, Electricity, and Magnetism

10	siemen per meter, S m^{-1}	millimho per centimeter, mmho cm^{-1}	0.1
10^4	tesla, T	gauss, G	10^{-4}

Water Measurement

9.73 × 10^{-3}	cubic meter, m^3	acre-inches, acre-in	102.8
9.81 × 10^{-3}	cubic meter per hour, m^3 h^{-1}	cubic feet per second, ft^3 s^{-1}	101.9
4.40	cubic meter per hour, m^3 h^{-1}	U.S. gallons per minute, gal min^{-1}	0.227
8.11	hectare-meters, ha-m	acre-feet, acre-ft	0.123
97.28	hectare-meters, ha-m	acre-inches, acre-in	1.03 × 10^{-2}
8.1 × 10^{-2}	hectare-centimeters, ha-cm	acre-feet, acre-ft	12.33

Concentrations

1	centimole per kilogram, cmol kg^{-1} (ion exchange capacity)	milliequivalents per 100 grams, meq 100 g^{-1}	1
0.1	gram per kilogram, g kg^{-1}	percent, %	10
1	milligram per kilogram, mg kg^{-1}	parts per million, ppm	1

Radioactivity

2.7 × 10^{-11}	bequerel, Bq	curie, Ci	3.7 × 10^{10}
2.7 × 10^{-2}	bequerel per kilogram, Bq kg^{-1}	picocurie per gram, pCi g^{-1}	37
100	gray, Gy (absorbed dose)	rad, rd	0.01
100	sievert, Sv (equivalent dose)	rem (roentgen equivalent man)	0.01

Plant Nutrient Conversion

	Elemental	Oxide	
2.29	P	P$_2$O$_5$	0.437
1.20	K	K$_2$O	0.830
1.39	Ca	CaO	0.715
1.66	Mg	MgO	0.602

1 Effects of Water on the Stability of Phospholipid Bilayers: The Problem of Imbibition Damage in Dry Organisms[1]

John H. Crowe, Lois M. Crowe, Folkert A. Hoekstra, and Christina Aurell Wistrom

University of California
Davis, California

When dry seeds are plunged into water, they imbibe water rapidly in the first few minutes, followed by a slower phase of imbibition until they become fully hydrated (Simon & Raja Harun, 1972; Leopold, 1980; Murphy & Noland, 1982). During the early stages of imbibition, the seeds leak solutes such as organic and inorganic ions, sugars, amino acids, and even proteins into the surrounding medium. This leakage, which has been unambiguously ascribed to loss of intracellular constituents (Simon & Raja Harun, 1972), often results in extensive embryo damage and even to its death (Simon, 1978). A similar phenomenon has also been reported in other organisms capable of surviving dehydration, including pollen (Hoekstra, 1984); anhydrobiotic nematodes (Crowe et al., 1979); and cysts of the brine shrimp, *Artemia* (Crowe et al., 1981). In this chapter, we provide a molecular explanation for this phenomenon.

The first clues concerning the mechanism of leakage came from anecdotal observations on conditions under which the leakage could be inhibited: If seed embryos or anhydrobiotic nematodes were imbibed to 0.25 to 0.3 g H_2O g^{-1} dry wt. (dw)[2] by placing them in moist air before they were placed in water, leakage was markedly slowed (Fig. 1-1). Further, dry but nonviable nematodes showed greatly increased leakage compared with anhydrobiotic ones. Leakage was nevertheless inhibited, even in the dead animals, by exposing them to moist air before they were plunged into water (Fig. 1-1). Similarly, heat-killed seed embryos showed significantly greater leakage during imbibition than did viable ones (Leopold, 1980). Thus, even though there are striking differences in leakage rates from viable and non-

---

[1] The work described here was conducted at the Univ. of California, Davis, CA 95616 and at the Agricultural Univ., Wageningen, Netherlands.
[2] To convert (g H_2O g^{-1} dw) to (%), multiply by 100. Thus, 0.10 g H_2O g^{-1} dw = 10%.

Copyright © 1989 Crop Science Society of America, 677 S. Segoe Rd., Madison, WI 53711, USA. *Seed Moisture*, CSSA Special Publication no. 14.

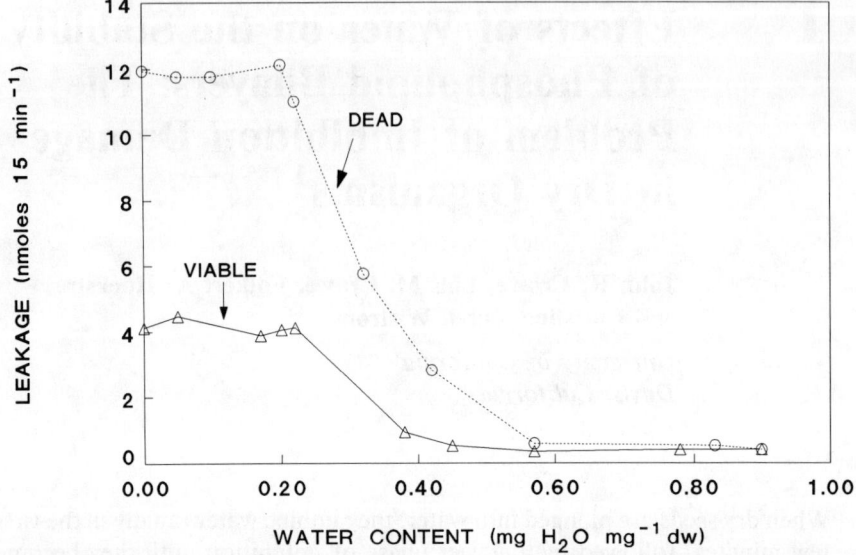

Fig. 1-1. Leakage of inorganic ions from dry nematodes as a function of their water content before they were plunged into water. The curve marked "dead" was produced from animals that had been rapidly dried, which kills them. The curve marked "viable" was produced from animals that had been slowly dried, during which anhydrobiosis is induced. (Redrawn from Crowe et al., 1979.)

viable dry organisms during imbibition, it appears that hydrating even the dead organisms from the vapor phase before they are plunged into water inhibits the leakage.

These results influenced Simon (1974, 1978) to hypothesize that the site of imbibition damage is the plasma membrane in seeds. He proposed that the damage could be explained based on the hydration-dependent phase behavior of membrane phospholipids. Specifically, he suggested that formation of nonbilayer phases [particularly the inverted hexagonal or hexagonal II (H_{II}) phase] in membrane phospholipids as a result of dehydration could result in leakage. The scientific community quickly embraced this interesting hypothesis, and numerous workers set out to seek evidence for the existence of H_{II} phase phospholipids in membranes of dry organisms. At this point, there is no convincing evidence that such nonbilayer phases are found in the membranes of *viable* dry organisms, as we will explain later. Therefore, little evidence exists to support Simon's hypothesis. We present here an alternative, but closely related hypothesis, that we believe will ultimately be the correct explanation for leakage during imbibition. We hasten to add at the outset, however, that while we do not believe Simon's (1974, 1978) suggestion is correct in detail, he provided the impetus for the studies described in this chapter. We, therefore, emphasize that the scientific community owes him a great debt for his important syntheses in the last decade that stimulated so many investigators, including ourselves, to study this phenomenon.

WHY DO CERTAIN LIPIDS FORM HEXAGONAL II PHASE?

Since the search for (H_{II}) phase lipids has played such a prominent role in this field, we wish first to describe briefly the mechanism by which some phospholipids enter this phase. At physiological temperatures and pressures and except under unusual conditions (e.g., Gruner, 1985), most phospholipids do not exhibit H_{II} phase. Phosphatidylethanolamine (PE) is the best-known commonly occurring phospholipid that enters this phase. It does so for the following reasons. The polar head group of PE possesses a positively charged N on the amino group that can form a bond of strong ionic character with the negatively charged phosphate of the adjacent PE. As a result, the polar head group occupies a small volume relative to that of the hydrocarbon chains in this molecule, leading to the following behavior. As the PE is heated, the hydrocarbon chains occupy an increasingly larger volume. Since the polar head groups interact so strongly through the amino-phosphoryl ionic bond, they are forced into a curved structure, leading to the formation of lipid tubes, with the polar head groups oriented toward the center of a hydrated core. Since increasing the unsaturation of the hydrocarbon chains will increase the volume they occupy, unsaturated PEs are more likely to enter the H_{II} phase than are saturated ones (Cullis & DeKruijff, 1979; Israelachvili et al., 1980). Similar phase behavior is found in cardiolipin in the presence of Ca^{2+} (Culliss & DeKruijff, 1979) and digalactosyl-containing lipids that are common in chloroplasts (Sen et al., 1981, 1983; Gounaris et al., 1983).

HOW MIGHT DEHYDRATION LEAD TO HEXAGONAL II PHASE?

Phospholipids are hydrated to some extent, with water molecules H bonded to the polar head group (reviewed in Crowe and Crowe, 1988a). This water plays a profound role in the phase behavior of phospholipids. In the case of PEs, for example, water removal from around the polar head group would be expected to decrease its volume relative to that of the hydrocarbon chains. As a consequence, dehydrated PEs enter H_{II} phase much more readily than do hydrated ones. In other words, the temperature at which the transition from lamellar to H_{II} phase (T_h) occurs is decreased (Seddon et al., 1983). In fact, there is good evidence, at least in principle, that any modifications to the head group that increase its volume relative to that of the hydrocarbon chains or which interrupt the bond between the amino N and phosphate of adjacent molecules will inhibit the lamellar-H_{II} phase transition (Silvius et al., 1986; Brown et al., 1986).

A further consequence of dehydration of PEs is that the gel-to-liquid crystalline transition temperature (T_m) in increased—opposite to the effect of dehydration on T_h. This effect is a further result of the decrease in the volume of the head group that accompanies dehydration; the increased opportunities for van der Waals' interactions among the hydrocarbon chains lead to an increased T_m. The net effect of the dehydration is that T_m and T_h tend to merge into a single transition (Seddon et al., 1983).

EFFECTS OF DEHYDRATION ON OTHER PHOSPHOLIPIDS

The phase behavior of other phospholipids that do not readily enter into H_{II} phase is also strongly affected by dehydration. For example, Chapman et al. (1967) found that T_m for dipalmitoylphosphatidylcholine (DPPC) is increased from about 41 °C in the presence of excess water to over 100 °C when the lipid is fully dehydrated. With the addition of a small amount of water to the dry lipid, T_m falls steadily, reaching a minimal value at about 10 to 12 mol H_2O mol^{-1} DPPC. Subsequent studies showed that addition of this amount of water increases the area occupied by each polar head group, thus decreasing van der Waals' interactions among the hydrocarbon chains (reviewed in Crowe & Crowe, 1988a).

DEHYDRATION AND LATERAL PHASE SEPARATIONS

Dehydration effects on phase behavior of membrane phospholipids can lead to lateral phase separations of membrane constituents in the plane of the membrane for the following reasons (see Crowe & Crowe, 1986a, 1988b; Crowe et al., 1987a, 1988a; ; Quinn, 1985 for a more complete discussion). The various phospholipid classes found in a typical native membrane possess various degrees of unsaturation in the hydrocarbon chains and varying capacities for binding water. As a result, during dehydration one might expect the various phospholipids to enter their respective gel phases at different times. The gel-phase domains would exclude more fluid domains, and as a consequence the dry membrane possesses phase-separated domains consisting of proteins, cholesterol, and the various phospholipid classes.

Lateral phase separation of the constituents of a membrane due to dehydration may have important consequences for a membrane with respect to Simon's H_{II} phase hypothesis; there is good evidence that PE is maintained in lamellar phase well above T_h in the presence of equimolar quantities of PC (Hui et al., 1981), an effect that is presumably due to inhibition of interactions between PE molecules by the presence of the PC (Hui et al., 1981). But if lateral phase separations of the PC and PE occur during dehydration, the result would be formation of pure domains of PE that presumably would then be capable of forming H_{II} phase. There is no evidence available concerning the phase behavior of simple, binary mixtures of PC and PE during dehydration, but studies on intact membranes suggest that lateral phase separations during dehydration do indeed lead to formation of H_{II} phase (Crowe & Crowe, 1982).

DO MEMBRANES OF ANHYDROBIOTIC ORGANISMS INCLUDE HEXAGONAL II LIPIDS?

Various workers have searched, with uniformly negative results, for H_{II} phase lipids in membranes or anhydrobiotic organisms, using several independent techniques.

1. X-ray diffraction—McKersie and Stinson (1980) and Sewaldt et al. (1981) have used x-ray diffraction to study both extracted phospholipids and seed particles and have found only lamellar phase at all water contents.
2. Nuclear magnetic resonance—Priestly and DeKruijff (1982) used ^{31}P-NMR to demonstrate that isolated phospholipids and membranes of dry pollen remained in lamellar phase even at the lowest water contents.
3. Freeze fracture—In our own laboratory, we have used freeze fracture to study membranes of anhydrobiotic nematodes and cysts of *Artemia*. The membranes of these dry organisms show no evidence of the presence of H_{II} phase lipids (reviewed in Crowe & Crowe, 1986b).

These results do not mean however, that H_{II} phase lipids are not present in membranes of organisms under stress. Indeed, there is excellent evidence that freezing (Steponkus, 1984) and dehydration (Crowe & Crowe, 1982) can lead to formation of H_{II} phase. But in such cases, the presence of this nonbilayer phase is indicative of such extensive damage that membrane function and viability of the whole organism are lost (Crowe et al., 1983).

HOW DO ANHYDROBIOTIC ORGANISMS ESCAPE FORMATION OF HEXAGONAL II PHASE?

It is still uncertain how anhydrobiotic organisms escape formation of H_{II} phase in their membrane lipids. We suggest at least the following possible mechanisms:

1. Such organisms lack significant quantities of PE or other H_{II} forming lipids in their membranes. Although the data are sparse, the existing information is inconsistent with this proposition (e.g., Womersley, 1981; Hoekstra, 1987, unpublished data).
2. Since in mixtures of PC and PE the PE is maintained in lamellar phase (Hui et al., 1981), it is possible that such a mechanism might prevent formation of H_{II} phase in membranes of anhydrobiotic organisms. In keeping with this suggestion is the finding of McKersie and Stinson (1980) that H_{II} phase is not formed in the polar lipid fraction isolated from seeds, even when the lipids were dried. This has been a puzzling result since one would expect phase separation of the PE and other lipids during drying, which would lead to H_{II} phase. But there have been no investigations of the phase behavior of even binary mixtures of PE and another lipid during drying other than the pioneering studies of Luzzatti (1968) on complex mixtures, so it is not possible to evaluate the effects of other lipids on PE in native membranes of anhydrobiotic organisms.
3. Recent evidence suggests that certain sugars profoundly affect the phase behavior of PE. These sugars include trehalose, commonly found at high concentrations in a wide variety of anhydrobiotic animals and lower plants (Crowe et al., 1984). There is a surprising paucity of evidence concerning the sugars present in, for example, seeds of higher plants (Leopold Vertucci, 1986), but the analog of trehalose in such organisms may be sucrose.

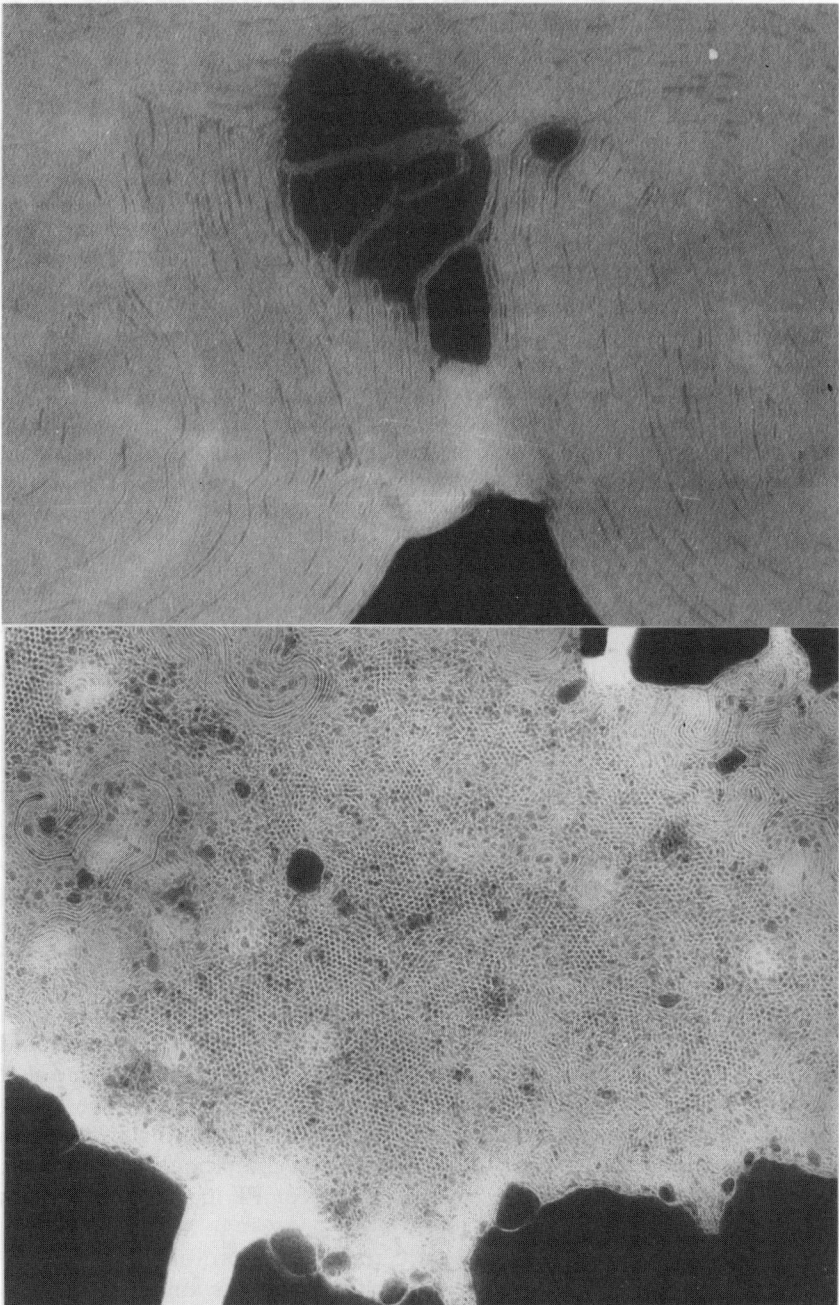

Fig. 1-2. Negative stained preparations of soybean PE dried in the presence (*top*) and absence (*bottom*) of trehalose. In the presence of trehalose the dry lipid is maintained in lamellae, while in the absence of the sugar the lipid forms hexagonally arranged tubes, ween here in end view.

Both these sugars may maintain dry PE in lamellar phase under conditions in which it is normally in H_{II} phase. For example, when soybean [*Glycine max* (L.) Merr.] PE is dried on an electron microscope grid and stained with phosphotungstic acid, H_{II} lipids can readily be seen as tubes, forming a close-packed honeycomb structure (Fig. 1-2). By contrast, when the same PE was dried in the presence of trehalose, it existed as lamellae (Fig. 1-2).

We suggest that the apparent inhibition of the lamellar-H_{II} transition is due to direct interaction between the sugar and the PE head group, probably by H bonding with the amino N and phsophate. This interaction would be expected to increase the apparent volume of the polar head group and to inhibit the phosphoryl-amino bonding required for formation of H_{II} phase (see Crowe & Crowe, 1988b for further discussion). We suspect that this is a major mechanism by which anhydrobiotic organisms escape formation of H_{II} phase in their membranes.

More recently, Goodrich et al., (1988) have shown similar effects on phase behavior or PE by sugars covalently linked to cholesterol in the bilayer via a hydrocarbon linkage. These workers suggested that the sugar residues may form H bonds with the PE head groups, similar to the mechanism described above.

TEMPERATURE EFFECTS ON LEAKAGE

Hoekstra (1984) showed that leakage from dry pollen was decreased by increasing the temperature at which the pollen was imbibed. Between 5 and 15 °C leakage sharply declined (Fig. 1-3). He also found that viability increased as leakage decreased (Fig. 1-3). These data strongly suggest that H_{II} phase lipids are not responsible for the leakage; increased temperature would *favor* formation of H_{II} phase but clearly inhibits the leakage (Fig. 1-3). It follows that H_{II} lipids cannot be responsible for the leakage, but there is an alternative mechanism that is entirely consistent with Hoekstra's data, discussed in the following section.

WHAT IS THE MECHANISM OF LEAKAGE?

We conclude from the previous discussion that the available evidence is not consistent with Simon's hypothesis that H_{II} phase is responsible for the leakage, and we now present an alternative suggestion. It is well known that as phospholipid bilayers pass through the gel-to-liquid, crystalline phase they increase in permeability (e.g., Hammoudah et al., 1981). Thus, it is possible that as pollen, for example, is rehydrated it passes through just such a phase transition, leading to leakage. What we are suggesting is that lipids in the plasma membranes of the pollen used in producing the data shown in Fig. 1-3 are in gel phase below 5 to 15 °C. When the pollen is heated above this temperature before it is imbibed, the lipids are in liquid crystalline phase,

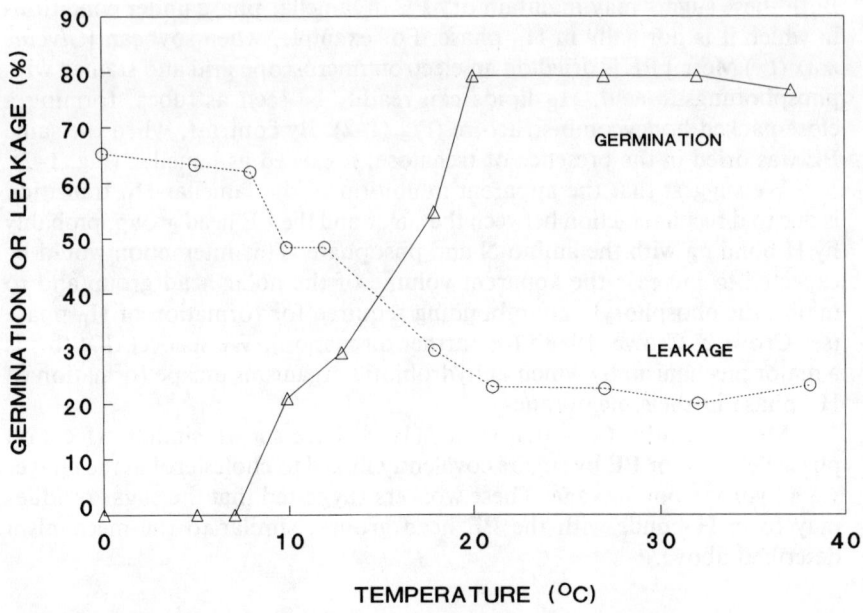

Fig. 1-3. Leakage of inorganic ions from and viability of *Typha* pollen as a function of the temperature at which the pollen was imbibed. (Redrawn from Hoekstra, 1984.)

even though they are dry. According to this hypothesis, upon rehydration, the membrane phospholipids in such pollen would not pass through the gel to liquid, crystalline phase and they would therefore not leak.

We have made progress towards testing this hypothesis, using infrared spectroscopy to record the gel-to-liquid, crystalline-phase transition. It is possible to detect CH_2 bands in infrared spectra of intact pollen grains of *Typha* carefully spread in a monolayer on the windows of the spectrometer (Fig. 1-4). The bands are broadened compared with those seen in pure phospholipids, but they are still clearly recognizable. As the pollen grains are heated, the CH_2 symmetric stretch increases in vibrational frequency between 10 and 15 °C (Fig. 1-4 and 1-5). This increase in frequency coincides with the rise in viability (Fig. 1-5). In fact, viability of the pollen following imbibition is a linear function of the vibrational frequency of the CH_2 bands (Fig. 1-6).

We interpret the data shown in Fig. 1-4 through 1-6 to be consistent with the hypothesis that below 10 °C gel-phase exist in the plasma membranes of these pollen grains and that the gel-to-liquid, crystalline transition that accompanies hydration is responsible for leakage. However, additional evidence will be required before we will be able to make this conclusion firmly. More recently, we have studied the phase properties of phospholipids and neutral lipids isolated from the pollen and have been able to assign the transition in the phospholipids with certainty (Crowe et al., 1988a). Further, we have found that the transition temperature for the phospholipids varies with water content, as expected (Crowe et al., 1988c).

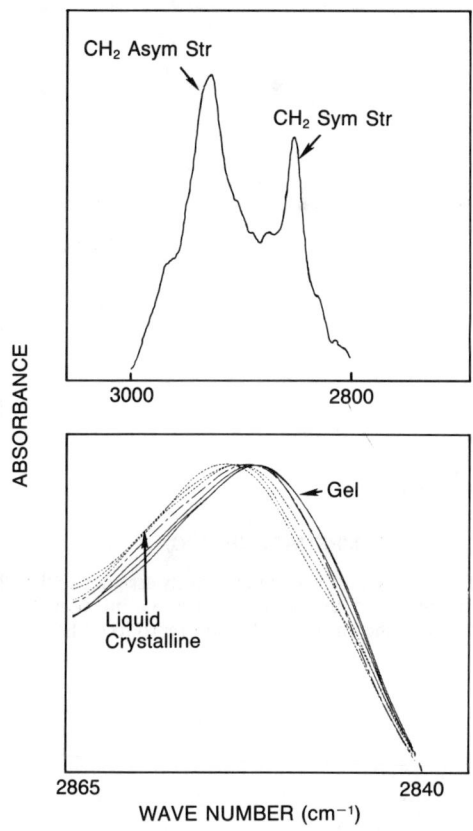

Fig. 1-4. *Top:* Infrared spectrum of inact pollen of *Typha* in the region of the assigned to the hydrocarbon chains in phospholipids. The major CH_2 and CH_3 vibrations are seen clearly. *Bottom:* Segments from spectra taken at different temperatures, showing shifts in frequency in the band tentatively assigned to the CH_2 symmetric stretch. The temperatures chosen are above and below the transition points for leakage and viability shown in Fig. 1-5.

WHY IS TRANSITION TEMPERATURE FOR DRY POLLEN SO LOW?

The apparent T_m for dry pollen is surprisingly low. In our first measurements of thermal transitions for dry phospholipids isolated from *Typha* pollen, we obtained an apparent T_m of about 70 °C—55 °C higher than the apparent T_m seen in the intact pollen. So we must ask how T_m is depressed in the intact pollen. The most likely hypothesis depends on the presence of high concentrations of sugars in the dry pollen. Hoekstra (1987, unpublished data) has found that *Typha* pollen contains about 25% of its dry weight in the form of sucrose. Both this molecule and trehalose have the ability to depress T_m of dry phospholipids effectively (Crowe et al., 1984, 1985). For example, hydrated DPPC has a T_m of 41 °C, and the dry lipid has a T_m of more than 100 °C (Chapman et al., 1967).

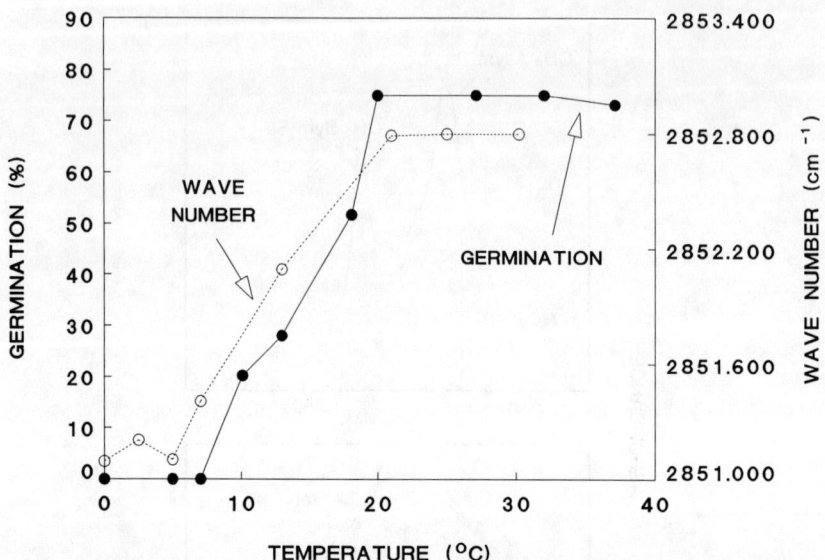

Fig. 1-5. Vibrational frequency for the band tentatively assigned to the CH_2 symmetric stretch in the spectra shown in Fig. 1-5 and similar spectra as a function of temperature. For comparison, the viability data shown in Fig. 1-4 are reproduced here.

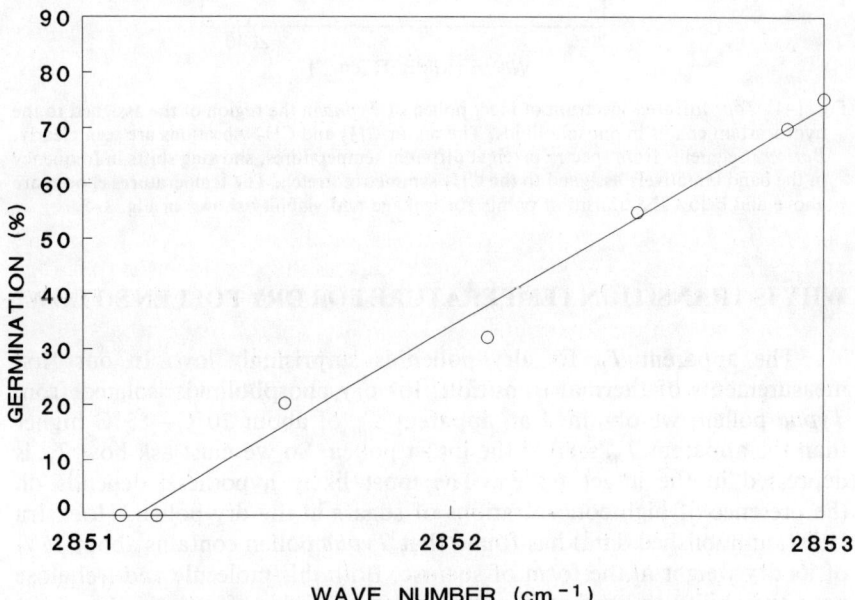

Fig. 1-6. Viability of pollen of *Typha* as a function of vibrational frequency for the CH_2 symmetric stretch at the time the pollen was imbibed.

When unilamellar vesicles of DPPC are dried in the presence of either sucrose or trehalose, a stable T_m of 24 °C is recorded—17 °C below T_m for the fully hydrated lipid (Crowe et al., 1984; Crowe & Crowe, 1988c). These results have subsequently been confirmed by several other laboratories (Strauss & Hauser, 1986; Lee et al., 1986). Thus, we suspect that T_m for phospholipids in the intact pollen is depressed by the presence of the sucrose. As a result, membrane phospholipids are maintained in liquid crystalline phase, even in the absence of water.

Depression of T_m is clearly a key event in preserving the structural integrity of the membrane. As a result of this effect, the lateral phase separations described earlier may be avoided completely since the membrane phospholipids may remain in liquid crystalline phase even in the absence of water. Furthermore, because T_m is depressed it makes it possible for phospholipids in *Typha* pollen, for example, to exist in liquid crystalline phase at room temperature even when dry. As a result, they do not pass from gel-to-liquid, crystalline phase during rehydration and they do not leak.

HOW DO SUGARS DEPRESS TRANSITION TEMPERATURE OF DRY PHOSPHOLIPIDS?

1. Retention of residual water by sugars: One possibility is that the sugars, which are highly hygroscopic, retain water in the dry mixtures. This water would, according to this hypothesis, be shared with phospholipids, so it would be the residual water that alters T_m, not the sugar itself. There is strong evidence that this is not the case; careful measurements show that dry phospholipid-sugar mixtures contain <0.1 mol H_2O mol^{-1} lipid (Crowe et al., 1987a).

2. Direct interaction between sugars and phospholipids: There is evidence from several laboratories now that there is a direct interaction between sugars and phospholipids (Anchordoguy et al., 1987; Crowe et al., 1984; Strauss & Hauser, 1986; Lee et al., 1986). Briefly, the evidence shows that there is H bonding between -OH groups on the sugar and the phosphate of the phospholipid. The sugar in effect replaces the water around the polar head group. Molecular modeling shows that in order to fit the sugar between adjacent head groups, expansion of the bilayer is required (Gaber et al., 1986). The result is that van der Waals' interactions among the hydrocarbon chains is decreased in the dry lipid, and T_m is consequently decreased.

CONCLUSION

We conclude from the available evidence that H_{II} phase lipids are not responsible for leakage from anhydrobiotic organisms during imbibition. Indeed, it is extremely doubtful that H_{II} phase lipids even occur in such organisms. We present what we believe to be the probable mechanism by which the organism escape formation of this deleterious nonbilayer phase.

We suggest that the leakage is due a gel-to-liquid, crystalline-phase transition during rehydration, and that the temperature at which this transition occurs is altered by the presence of sugars, particularly sucrose and trehalose, in the dry organisms. Finally, we wish to emphasize that it is now feasible to produce hydration-dependent phase diagrams for membrane phospholipids of commercial seeds, pollen, and the like, using infrared spectroscopy and other available methods such as high-sensitivity differential scanning calorimetry. With such a phase diagram in hand, it would then be possible to determine the combination of temperature and water content to produce optimal survival, an application of the basic research described here that has clear, immediate relevance to agriculture.

ACKNOWLEDGMENT

We gratefully acknowledge support of the National Science Foundation through grants PCM 83-17538 and DMB 85-18194.

REFERENCES

Anchordoguy, T.J., A.S. Rudolph, J.F. Carpenter, and J.H. Crowe. 1987. Modes of interaction of cryoprotectants with membrane phospholipids during freezing. Cryobiology 24:324–331.

Brown, P.M., J. Steers, S.W. Hui, P.L. Yeagle, and J.R. Silvius. 1986. Role of head group structure in the phase behavior of amino phospholipids. II. Lemellar and non-lamellar phases of unsaturated phosphatidylethanolamine analogues. Biochemistry 25:4259–4267.

Chapman, D., R.M. Williams, and B.D. Ladbrooke. 1967. Physical studies of phospholipids. VI. Thermotropic and lyotropic mesomorphism of some 1,2-diacyl-phosphatidylcholines (lecithens). Chem. Phys. Lipids 1:445–475.

Crowe, J.H., and L.M. Crowe. 1986a. Stabilization of membranes, metabolism, and dry organisms. Cornell Univ. Press, Ithaca, NY.

----, and ----. 1988a. Effects of water and sugar on the stability of phospholipid bilayers. In G. Benga (ed.) Water transport in biological membranes. CRC Press, Boca Raton, FL. (In press.)

----, ----, J.F. Carpenter, and C. Aurell Wistrom. 1987a. Stabilization of dry phospholipid bilayers and proteins by sugars. Biochem. J. 242:1–10.

----, ----, ----, A.S. Rudolph, C. Aurell Wistrom, B.J. Spargo, and T.J. Anchordoguy. 1988a. Interactions of sugars with membranes. Biochim. Biophys. Acta Rev. (In press.)

----, ----, and D. Chapman. 1984. Preservation of membranes in anhydrobiotic organisms: The role of trehalose. Science 223:701–703.

----, ----, and S.A. Jackson. 1983. Preservation of structural and functional activity in lyophilized sarcoplasmic reticulum. Arch. Biochem. Biophys. 220:477–484.

----, ----, and S.J. O'Dell. 1981. Ice formation during freezing of Artemia cysts of variable water contents. Mol. Physiol. 1:145–152.

----, F.A. Hoekstra, and L.M. Crowe. 1988b. Gel to liquid crystalline phase transitions are responsible for imbibitional leakage in dry pollen. Proc. Natl. Acad. Sci., USA (in press).

----, ----, ----, T.J. Anchordoguy, and E. Drobnis. 1988c. Lipid phase transitions measured in intact cells with Fourier transform infrared spectroscopy. Cryobiology (in press).

----, S.J. O'Dell, and D.A. Armstrong. 1979. Anhydrobiosis in nematodes: Permeability during hydration. J. Exp. Zool. 207:431–438.

----, B.J. Spargo, and L.M. Crowe. 1987b. Preservation of dry liposomes does not require retention of residual water. Proc. Natl. Acad. Sci. USA 84:1537–1540.

Crowe, L.M., and J.H. Crowe. 1982. Hydration-dependent hexagonal phase in biological membrane. Arch. Biochem. Biophys. 217:582–587.

----, and ----. 1986b. Hydration-dependent phase transitions and permeability properties of biological membranes. p. 210–230. In A.C. Leopold (ed.) Membranes, metabolism, and dry organisms. Cornell Univ. Press, Ithaca, NY.

----, and ----. 1987b. Effects of water and carbohydrates on membrane fluidity. In R.C. Aloia (ed.) Advances in membrane fluidity. (In press.)

----, and ----. 1988b. Trehalose and dry dipalmitoylphosphatidylcholine revisited. Biochim. Biophys. Acta (in press).

----, ----, and D. Chapman. 1985. Interaction of carbohydrates with dry dipalmitoylphosphatidylcholine. Arch. Biochem. Biophys. 236:289–296.

Cullis, P.R., and B. DeKruijff. 1979. Polymorphism and the role of lipids in biological membranes. Biochim. Biophys. Acta 559:399–420.

Gaber, B.P., I. Chandrasekhar, and N. Pattabiraman. 1986. The interaction of trehalose with the phospholipid bilayer: A molecular modeling study. p. 231–241. In A.C. Leopold (ed.) Membranes, metabolism, and dry organisms. Cornell Univ. Press, Ithaca, NY.

Goodrich, R.P., T.M. Handel, J. Baldeschwieler. 1988. Modification of lipid phase behavior with membrane bound cryoprotectants. Biochim. Biophys. Acta. (In press.)

Gounaris, K., D.A. Mannock, A. Sen., A.P.R. Brain, W.P. Williams, and P.J. Quinn. 1983. Polyunsaturated fatty acid residues of galactolipids are involved in the control of bilayer/nonbilayer lipid transitions in higher plant chloroplasts. Biochim. Biophys. Acta 732:229–242.

Gruner, S.M. 1985. Intrinsic curvature phyothesis for biomembrane lipid composition: A role for non-bilayer lipids. Proc. Natl. Acad. Sci. USA 82:3665–3669.

Hammoudah, M.M., S. Nir, J. Bentz, E. Mayhew, T.P. Stewart, S.W. Hui, and R.J. Kurian. 1981. Interactions of La with phosphatidylserine vesicles. Binding, phase transition, leakage, ^{31}P-NMR and fusion. Biochim. Biophys. Acta 645:102–114.

Hoekstra, F.A. 1984. Imbibitional chilling injury in pollen: Involvement of the respiratory chain. Plant Physiol. 74:815–821.

Hui, S.W., T.P. Stewart, P.L. Yeagle, and A.D. Albert. 1981. Bilayer to nonbilayer transitions in mixtures of phosphatidylethanolamine and phosphatidylcholine: Implications for membrane properties. Biochim. Biophys. Acta 207:227–240.

Israelachvili, J.N., S. Marcelja, and R.G. Horn. 1980. Physical principles of membrane organization. Q. Rev. Biophys. 13:121–200.

Lee, C.W.B., J.S. Waugh, and R.G. Griffin. 1986. Solid state NMR study of trehalose/1,2-dipalmitoyl-sn-phosphatidylcholine interactions. Biochemistry 25:3737–3742.

Leopold, A.C. 1980. Temperature effects on soybean imbibition and leakage. Plant Physiol. 65:1096–1102.

----, and C.W. Verticci. 1986. Physical attributes of desiccated seeds. p,. 22–34. In A.C. Leopold (ed.) Membranes, metabolism, and dry organisms. Cornell Univ. Press, Ithaca, NY.

Luzzati, V. 1968. X-ray diffraction studies of lipid-water systems in biological membranes. p. 71–124. In D. Chapman (ed.) Biological membranes. Vol. I. Academic Press, New York.

McKersie, B.D., and R.H. Stinson. 1980. Effect of dehydration on leakage and membrane structure in Lotus corniculatus L. seeds. Plant Physiol. 66:316–320.

Murphy, J.B., and T.L. Noland. 1982. Temperature effects on seed imbibition and leakage mediated by viscosity and membranes. Plant Physiol. 69:428–435.

Priestly, D.A., and B. DeKruijff. 1982. Phospholipid motional characteristics in a dry biology system. Plant Physiol. 70:1075–1083.

Quinn, P.J. 1985. A lipid phase separation model of low-temperature damage to biological membranes. Cryobiology 22:128–147.

Seddon, J.M., G. Cevc, and D. Marsh. 1983. Calorimetric studies of the gel-fluid (L_β-L_α) and lamellar-inverted hexagonal (L_α-H_{II}) phase transitions in dialkyl- and diacylphosphatidylethanolamines. Biochemistry 22:1280–1289.

Sen, A., D.A. Mannock, D.J. Collin, P.J. Quinn, and R.P. Williams. 1983. Thermotropic phase properties and structure of 1,2-distearoylgalactosylglycerols in aqueous systems. Proc. R. Soc. London B 218:349–364.

----, P. Williams, and P.J. Quinn. 1981. The structure and thermotropic properties of pure 1,2-diacylgalactosylglycerols in aqueous systems. Biochim. Biophys. Acta 663:380–389.

Sewaldt, V., D.A. Priestley, A.C. Leopold, G.W. Feigenson, and F. Goodsaid-Zaldvondo. 1981. Membrane organization in soybean seeds during hydration. Planta 152:19–23.

Silvius, J.R., P.M. Brown, T.J. O'Leary. 1986. Role of headgroup structure in the phase behavior of amino phospholipids. I. Hydrated and dehydrated lamellar phases of saturated phosphatidylethanolamine analogues. Biochemistry 25:4249-4258.

Simon, E.W. 1974. Phospholipids and plant membrane permeability. New Phytol. 73:377-420.

----, and R.M. Raja Harun. 1972. Leakage during seed imbibition. J. Exp. Bot. 23:1076-1085.

----. 1978. Membranes in dry and imbibing seeds. p. 205-224. In J.H. Crowe and J.S. Clegg (ed.) Dry biological systems. Academic Press, New York.

Steponkus, P.L. 1984. Role of the plasma membrane in freezing injury and cold acclimation. Ann. Rev. Plant Physiol. 35:443-484.

Strauss, G., and H. Hauser. 1986. Stabilization of lipid bilayer vesicles by sucrose during freezing. Proc. Natl. Acad. Sci., USA 83:2422-2426.

Womersley, C. 1981. Biochemical and physiological aspects of anhydrobiosis. Comp. Biochem. Physiol. 70B:669-678.

2 Seed Moisture: Recalcitrant vs. Orthodox Seeds[1]

H. F. Chin and B. Krishnapillay
Universiti Pertanian Malaysia
Selangor, Malaysia

P. C. Stanwood
USDA-ARS
Fort Collins, Colorado

The terms *orthodox* and *recalcitrant seeds* came into usage in 1973. Roberts (1973) classified these seeds according to their physiological behavior. Orthodox seeds were those seeds that could be dried to low moisture content (e.g., 0.05 g H_2O g^{-1} fresh wt., fw)[2] and tolerated freezing temperatures. Harrington's rule of thumb (1972) states that between 0.05 and 0.14 g H_2O g^{-1} fw if the seed moisture content is increased by 1% (0.01 g H_2O g^{-1} fw), the life span of orthodox seeds will be reduced by 50%. Recalcitrant seeds could not be dried below a relatively critical moisture content (e.g., 0.30 g H_2O g^{-1} fw) and could not tolerate freezing temperatures. Because of these differences, Hanson (1984) suggested that the term *desiccation sensitive* more accurately described recalcitrant seeds. Recalcitrant seeds lose viability once they are dried to a moisture content below a relatively high critical value. This means that seed moisture is a critical factor determining the viability and longevity of both recalcitrant and orthodox seeds.

For this reason, one must first identify the seed type before prescribing a method of storage. Orthodox seeds require low (e.g., 0.05 g H_2O g^{-1}fw) seed moisture content for successful long-term storage. In contrast, recalcitrant seeds must be stored at relatively high moisture levels (e.g., 0.30 g H_2O g^{-1} fw). As a result, the methods, techniques, packaging materials, containers, and the storage environment must be modified for successful storage of these seeds. In this chapter, determination of recalcitrant seeds moisture content is described and discussed. In addition, differences between recalcitrant and orthodox seed structure, morphology, and physiology are reviewed with particular emphasis on seed moisture content.

---

[1] Contributions from the Agronomy and Horticulture Dep., Universiti Pertanian Malaysia, 43400 Serdang, Selangor, Malaysia and National Seed Storage Laboratory, USDA-ARS, Colorado State Univ., Fort Collins, CO 80523, USA.
[2] To convert (g H_2O g^{-1} fw) to (%), multiply by 100. Thus, 0.10 g H_2O g^{-1} fw = 10%.

Copyright © 1989 Crop Science Society of America, 677 S. Segoe Rd., Madison, WI 53711, USA. *Seed Moisture*, CSSA Special Publication no. 14.

MOISTURE DETERMINATION

The International Seed Testing Association (ISTA) has published rules for the determination of orthodox seed moisture content, which have changed considerably from 1931 to 1985 (Grabe, 1987). In the latest version (ISTA, 1985), two 5.0-g samples of seeds are used for the determination of seed moisture content. High and low temperature air-oven methods of 130 to 133 °C and 101 to 103 °C are recommended for non-oily and oily seeds, respectively. If seeds are large, e.g., maize (*Zea mays* L.), ground samples are required. The duration of the test for the high-temperature method is 1 to 4 h; for the low-temperature method, the duration is 16 to 18 h.

For tree seeds, ISTA (1966) recommended the air-oven method at 101 to 105 °C except for *Abies, Cedrus, Fagus, Picea,* and *Tsuga* for which the toluene-distillation method could be used. Bonner (1972) concluded that the air-oven method at 130 °C for 4 h followed by 2 h of cooling in a desiccator provided accurate results for seeds of such North American hardwoods as *Plantanus occidentalis* L., *Liquidambar styracifona* L., and *Fraxinus pennsylvanica* Marsh.

However, there are still no rules for drying methods for recalcitrant seeds. Since those seeds are typically large and heavy, 5.0-g drying samples are not adequate in size, which makes sampling difficult. The number of seeds that should be tested to obtain a statistically reliable drying method has yet to be determined. It should also be emphasized that most recalcitrant seeds are sold by number and not by weight. Hence, the economics of seed number for testing and the number of seeds available for an experiment should be kept in mind. According to Cochran (1953) and Mok (1972), the optimum sample size for large recalcitrant seeds should be approximately 20 seeds. Once the sample size is obtained, the typical method for determining moisture content of recalcitrant seeds is to cut 1.0-mm seed cross sections. These sections are placed in an oven at 101 to 105 °C for 16 h. Using whole seeds, halving or quartering the seeds produces more variable seed moisture results.

DIFFERENCES BETWEEN RECALCITRANT AND ORTHODOX SEEDS

Recalcitrant and orthodox seeds differ greatly in their ecology and morphology. Recalcitrant seeds are primarily from perennial trees in the moist tropics. In some cases, they also come from temperate tree or aquatic species, while most orthodox seeds are from annual species grown in open fields.

With respect to morphology, recalcitrant seeds differ from orthodox seeds not only in size but also complexity and viability. Many recalcitrant seeds for example, are not true seeds but exist as fruits, an example being a fibrous drupe in the case of coconuts (*Cocos nucifera* L.). Generally, recalcitrant seeds are covered with fleshy or juicy arilloid layers and impermeable testa. These maternal structures maintain the seeds in a high-moisture environment. Recalcitrant seeds also have greater variation in the shape and size of cotyledons and the embryonic attachment compared to bean (*Phaseolus vulgaris* L.), a typical example of an orthodox seed.

Moisture Content of Recalcitrant and Orthodox Seeds

At physiological maturity, the moisture contents of recalcitrant seeds (0.50-0.70 g H_2O g^{-1} fw) are much higher than orthodox seeds (0.30-0.50 g H_2O g^{-1} fw). At the same time, recalcitrant seeds are larger (42 × 25 mm) and heavier, an average of 14.0 g per seed for durian (*Durio zibethinus* Murr.). Typical differences in seed size, 1000 seed weight, and moisture content of recalcitrant and orthodox seeds are presented in Table 2-1. Orthodox seeds undergo drying after physiological maturity: initially, the moisture content is high (0.30-0.50 g H_2O g^{-1}fw), but they then will dry to a harvest moisture content of 0.15 to 0.20 g H_2O g^{-1} fw).

Although recalcitrant seeds are large, their embryos in relation to the whole seed are only about 15% of the size of those for orthodox seeds. Typical examples of the proportion of the total seed dry weight for the cotyledons, testa and embryos of recalcitrant and orthodox seeds are given in Table 2-2.

Table 2-1. Seed size, 1000 seed wt., and moisture content of typical recalcitrant and orthodox seeds.

Crop species	Seed size, length × width	1000 Seed wt.	Moisture content
	mm	g	g H_2O g^{-1} fw
Recalcitrant			
Nephelium lappaceum L.	28 × 16	3 555	0.49
Artocarpus heterophyllus Lam.	35 × 24	8 520	0.52
Artocarpus champeden (Lour.) Spreng.	30 × 20	5 814	0.71
Lansium domesticum Corr.	17 × 13	2 335	0.52
Bouea ganadaria	22 × 15	3 530	0.46
Durio zibethinus Murr.	42 × 25	14 783	0.50
Theobroma cacao L.	25 × 25	1 995	0.36
Orthodox			
Hibiscus esculentus L.	6 × 4	146	0.18
Vigna sesquipedalis (L.) Fruw.	12 × 5	192	0.16

Table 2-2. The percentage dry weight for the cotyledons, testa, and embryo of orthodox and recalcitrant seeds.

Crop species	Cotyledon	Testa	Embryo
		% dry wt.	
Orthodox			
Pisum sativum L.	86.7	11.9	1.4
Phaseolus vulgaris L.	90.4	8.3	1.3
Glycine max (L.) Merr.	90.6	7.3	2.1
Vigna sesquipedalis (L.) Fruw.	83.3	15.8	0.9
Hibiscus esculentus L.	82.3	16.4	1.3
Recalcitrant			
Artocarpus heterophyllus Lam.	91.1	8.7	0.2
Nephelium lappaceum L.	88.5	11.2	0.3

The distribution of moisture content in the embryo and whole seed for recalcitrant and orthodox seeds are shown in Table 2-3. The greatest difference in moisture content is found in the variation between whole seeds and embryos of recalcitrant seeds. Even greater variations in the embryos for both seed types are depicted by the higher coefficients of variation (CVs) (Table 2-3). Whole seed moisture content of orthodox seeds are constant as exemplified by the CV of 2.8% that is in contrast to recalcitrant seeds that have a CV above 7.0%. Embryos also possess greater moisture contents when compared to whole seeds. They are easily dried to a lower moisture content, however, because of their smaller size. In the case of recalcitrant seeds, embryos account for only 0.25% of the dry weight of whole seed compared to an average of 1.4% for orthodox seeds.

Recalcitrant seeds, because of their differences and variation in size and moisture content, do not undergo similar maturation drying processes as orthodox seeds. The great variation in moisture content between individual seeds makes a seed lot heterogeneous and, consequently, experimental results can be erratic with variable germination values compared to orthodox seeds. Researchers working on recalcitrant seed storage often encounter unexplained results that are probably attributable to the greater variation in seed traits between individual recalcitrant seeds.

THE IMPORTANCE AND PROBLEMS OF SEED MOISTURE IN RECALCITRANT SEEDS

High moisture contents of recalcitrant seeds makes them sensitive to desiccation and chilling injury. For example, cocoa (*Theobroma cacao*) and *Hevea* seeds rapidly lose germination when they are dried to 0.26 and 0.20 g H_2O g^{-1} fw, respectively (Chin et al., 1981; Hor et al., 1984). These values are extremely high compared to orthodox species that can be safely dried to 0.02 to 0.04 g H_2O g^{-1} fw. According to King and Roberts (1980), the large seed and impermeable seedcoat traits benefit recalcitrant seeds since

Table 2-3. Variations in moisture content of embryos and whole seeds for recalcitrant and orthodox seeds with their coefficients of variation.

Crop	Moisture content Embryo	Whole seed	Coefficient of variation Embryo	Whole seed
	g H_2O g^{-1} fw		%	
Recalcitrant				
Nephelium lappaceum L.	0.334	0.490	27.1	13.4
Artocarpus heterophyllus Lam.	0.396	0.522	12.3	52.9
Artocarpus champeden (Lour.) Spring.	0.749	0.707	12.3	48.7
Lansium domesticum Corr.	0.493	0.517	29.8	94.7
Bouea ganadaria	0.483	0.463	21.2	5.9
Durio zibethinus Murr.	0.652	0.499	9.5	9.4
Theobroma cacao L.	0.662	0.362	9.7	7.8
Orthodox				
Hibiscus esculentus L.	0.291	0.184	6.7	2.9
Vigna sesquipedalis (L.) Fruw.	0.126	0.157	3.8	2.7

they are less likely to be affected by minor fluctuations in relative humidities that might occur prior to germination. Because of their large seed size, recalcitrant seeds rely on plasmodesmata for intercellular transport that can be disrupted by drying and lead to subsequent viability loss (Livingston, 1964). This is one reason why the identification of recalcitrant seeds is important since a number of recalcitrant species have now been reclassified as orthodox, e.g., citrus (*Citrus limon* L.) (Mumford & Grout, 1979) and cassava (*Manihot esculenta* Crantz) (Ellis et al., 1981). Therefore, the correct diagnosis and identification are of prime importance in describing recalcitrant seeds since improper storage conditions can result in total viability loss. Chin et al. (1987) have recommended a low-temperature drying protocol using a desiccant to minimize deterioration of stored recalcitrant seeds.

The exact cause of recalcitrant seed death and its relationship with moisture content is not fully understood or investigated. Seed death could be due either to the moisture content falling below a certain critical value or simply a general physiological deterioration with time. If viability loss due to drying below a certain critical value is the prime mechanism governing deterioration, then the practical approach is to not dry the seeds below that critical value but store them in a subimbibed state. In contrast, the loss of seed viability due to physiological deterioration with time in storage should mimic the gradual physiological deterioration observed for orthodox seeds. In this case, it is accumulated damage as a result of potentially impaired mechanisms (Villiers, 1972). It is also possible that drying has other effects such as alterations of enzyme structure, degradation of cell membranes (Chin et al., 1984), and release of phenolic compounds leading to the loss of enzyme activity (Loomis & Battaile, 1966). All of these mechanisms to explain desiccation injury and viability loss in recalcitrant seeds require further investigation before a successful method for the long-term storage of recalcitrant seeds is developed.

High seed moisture content is associated with freezing injury through ice crystal formation that disrupts cells when subjected to subzero temperatures. However, the deleterious effects of temperature below 16 to 18 °C (chilling injury) on a number of recalcitrant seeds has been observed and is difficult to explain in terms of freezing damage. Species susceptible to this temperature are cocoa (Hor, 1984), *Shorea ovalis* (Sasaki, 1976), and *Drybalanops aromatica* (Jensen, 1971). The loss in viability of cocoa seeds is abrupt, e.g., a drop from 17 to 15 °C kills the seeds. Boroughs and Hunter (1963) suggested three possible reasons for this rapid decline in germination of recalcitrant seeds due to low temperatures: (i) the presence of a temperature dependent, rate-limiting reaction, the cessation that causes lethal metabolic disruption; (ii) the absence of a protective substance in seeds not susceptible to chilling; and (iii) the liberation of a toxic material due to cold-induced changes in mambrane permeability.

Imbibition is a prerequiste for germination for both orthodox and recalcitrant seeds. The only difference with recalcitrant seeds is that they tend to be fully imbibed in the fruit resulting in a number of seeds germinating within the fruit. In the case of jackfruit (*Artocarpus heterophyllus* Lam.),

80% of the seeds can germinate inside the ripe fruit. Farrant et al. (1986) attributed this early germination to the advanced maturity of recalcitrant seeds. This emphasizes problems of storing recalcitrant seeds since these seeds are commonly maintained in conditions similar to full imbibition. One technique commonly used is to reduce storage temperature to approximately 5 °C, if the seeds can tolerate it. Alternatively, germination inhibitors may also be used. For example, natural inhibitors occur in fruits of mangosteen (*Garcinia mangostana* L.) (Winters & Rodriquez-Colon, 1953) and rambutan (*Nephelium lappaceum* L.) (Chin, 1975). Chemical inhibitors also have been used but have proven uneffective (Goldbach, 1979).

Finally, high moisture in recalcitrant and orthodox seeds is a problem for successful seed storage due to its association with microbial contamination. In general, even for orthodox seeds, when the seed moisture content is in excess of 0.10 to 0.13 g H_2O g^{-1} fw, fungal invasions rapidly destroy seed viability (Harrington, 1963). This is particularly true in the wet, humid tropics where the relative humidity and temperature are always high. As a result, microbial contamination is a serious problem with the storage of recalcitrant seeds, where the imbibed state is the only practical storage method available today. A partial drying method has been proposed that can prolong the storage life of recalcitrant seeds. For example, partial drying of cocoa seeds followed by a fungicide treatment of thiram mixture and storing in air-conditioned rooms at 20 °C has prolonged the storage life of these seeds for 24 wk (Hor, 1984).

FUTURE RESEARCH

Roberts et al. (1984) have suggested that the most promising method of germplasm preservation for recalcitrant species may be storage in liquid nitrogen. Stanwood (1983) listed a moisture range from a low of 0.096 g H_2O g^{-1} fw for sesame (*Sesamum indicum* L.) to 0.285 g H_2O g^{-1} fw for bean as the high moisture freezing limit (HMFL). So, it is possible that recalcitrant seeds or their embryos can be dried to the higher HMFL limits and subsequently frozen in liquid nitrogen. In 1986, the International Board for Plant Genetic Resources (IBPGR) funded a project on dehydration and preservation techniques of recalcitrant seeds at the Universiti Pertanian Malaysia in collaboration with National Seed Storage Laboratory, Fort Collins, CO. To date, embryos of a few recalcitrant species (*Artocarpus heterophyllus, Nephelium lappaceum, Cocos nucifera,* and *Drybalanops aromatica*) have survived cryopreservation. Culturing these species in enriched media showed signs of growth, callus, shoots, roots, and seedlings (Chin et al., 1987). Rubber (*Hevea brasiliensis*), a true recalcitrant species, has embryos that, after partial drying, have been successfully cryopreserved for the first time (Normah et al., 1986). These techniques have to be refined, however, and they may differ from species to species. The technique of in vitro storage has advantages and disadvantages. It may have to be used in the future if no progress can be made in a particular species and all other storage methods have failed. The in vitro technique using cryopreserved embryos may prove successful in the future considering the rapid advances being made in the field of biotechnology.

CONCLUSION

Improvement in seed science technology has advanced rapidly over the last century in various aspects of orthodox seeds; yet, the desiccation-sensitive, recalcitrant seeds remain an enigma. We still have few successful techniques to store these seeds for more than 26 wk and the accurate determination of their moisture content remains unresolved as evidence by the lack of rules prescribed by ISTA.

Basic studies on the identification of the causes of desiccation sensitivity in recalcitrant seeds and their intolerance to low temperatures of 10 to 15 °C should also be given high priority. These are important factors since any long-term storage of recalcitrant seeds requires preventive measures for seed deterioration and viability loss. Alternatives to imbibed seed storage, such as partial drying or lowering moisture contents to just above a critical level in conjunction with fungicide treatments, also appear promising.

REFERENCES

Bonner, F.T. 1972. Measurement of moisture content in seeds of some American hardwoods. Proc. Int. Seed Test. Assoc. 37:975-983.
Boroughs, H., and J.R. Hunter. 1963. The effect of temperature on the germination of cocoa seeds. Proc. Am. Soc. Hortic. Sci. 82:222.
Chin, H.F. 1975. Germination and storage of rambutan (*Nephelium lappaceum*) seeds. Malays. Agric. Res. 4:173-180.
----, M. Aziz, B.B., and S. Hamzah. 1981. The effect of moisture and temperature on the ultrastructure and viability of seeds of *Hevea brasiliensis*. Seed Sci. Technol. 9:411-422.
----, Y.L. Hor, and M.B. Mohd. Lassim. 1984. Identification of recalcitrant seeds. Seed Sci. Technol. 12:429-436.
----, ----, and B. Krishnapillay. 1987. Dehydration and preservation techniques of recalcitrant seeds. Second Progress Rep. Int. Board for Plant Genet. Resources, June 1987. Int. Board for Plant Genet. Resources, Rome.
Cochran, W.G. 1953. Sampling techniques. John Wiley and Sons, New York.
Ellis, R.H., T.D. Hong, and E.H. Roberts. 1981. The influence of desiccation on cassava seed germination and longevity. Ann. Bot. 47:173-175.
Farrant, J.M., N.W. Pammenter, and P. Berjak. 1986. Recalcitrance—A current assessment. Preprint 15. Int. Seed Test. Assoc. Congr., Brisbane, Australia.
Grabe, D.F. 1987. Report of the seed moisture committee 1983-1986. Seed Sci. Technol. 15:451-452.
Goldbach, H. 1979. Imbibed storage of *Melicoccus bijugatus* and *Eugenia brasiliensis* (*E. dombeyi*) using abscisic acid as a germination inhibitor. Seed Sci. Technol. 7:403-406.
Hanson, J. 1984. The storage of seeds of tropical tree fruits. p. 53-62. I*v* J.H.W. Holden and J.T. Williams (ed.) Crop genetic resources: Conservation and evaluation. Allen and Unwin, London.
Harrington, J.F. 1963. Practical advice and instructions on seed storage. Proc. Int. Seed Test. Assoc. 28:989-994.
----. 1972. Seed storage and longevity. p. 145-250. *In* T.I. Kozlowski (ed.) Seed biology. III. Insects and seed collection, storage and seed testing. Academic Press, New York.

Hor, Y.L. 1984. Storage of cocoa (*Theobroma cacao*) seeds and changes associated with their deterioration. Ph.D. diss. Universiti Pertanian Malaysia.

----, H.F. Chin, and Zain Karim, Mohd. 1984. The effect of seed moisture and storage temperature on the storability of cocoa (*Theobroma cacao*) seeds. Seed Sci. Technol. 12:415-420.

International Seed Testing Association. 1966. International rules for seed testing. Proc. Int. Seed Test. Assoc. 31:128-134.

----. 1985. International rules for seed testing. Seed Sci. Technol. 13:338-341, 493-495.

Jensen, L.A. 1971. Observations on the viability of Borneo camphor (*Drybalanops aromatica* Gaertn.). Proc. Int. Seed Test. Assoc. 36:141-146.

King, K.W., and E.H. Roberts. 1980. Maintenance of recalcitrant seeds. Achievement and possible approaches. A report on a literature review carried out for the Int. Board for Plant Genetic. Resources. Int. Board for Plant Genet. Resources, Rome.

Livingston, L.G. 1964. The nature of plasmodesmata in normal (living) plant tissues. Am. J. Bot. 51:950-957.

Loomis, W.S., and J. Battaile. 1966. Plant phenolic compounds and the isolation of plant enzymes. Phytochemistry 5:423-438.

Mok, C.K. 1972. Sample size for moisture and viability testing of oil palm (*Elaeis guineensis* Jacq.) seeds. Proc. Int. Seed Test. Assoc. 37:751-761.

Mumford, P.M., and B.W.W. Grout. 1979. Desiccation and low temperature (−196 C) tolerance of *Citrus limon* seed. Seed Sci. Technol. 7:407-410.

Normah, M.N., H.F. Chin, and Y.L. Hor. 1986. Desiccation and cryopreservation of embryonic axes of *Hevea brasiliensis* Muell-Arg. Pertanika 9:299-303.

Roberts, E.H. 1973. Predicting the storage life of seeds. Seed Sci. Technol. 1:499-514.

----, M.W. King, and R.H. Ellis. 1984. Recalcitrant seeds: Their recognition and storage. p. 38-52. *In* J.H.W. Holden and J.T. Williams (ed.) Crop genetic resources: Conservation and evaluation. Allen and Unwin, London.

Sasaki, S. 1976. The physiology, storage and germination of timber seeds. p. 11-15. *In* H.F. Chin et al. (ed.) Seed technology in the tropics. Universiti Pertanian Malaysia, Serdang, Selangor, Malaysia.

Stanwood, P.C. 1983. Cryopreservation of seed germplasm for genetic conservation. p. 200-226. *In* K.K. Kartha (ed.) Cryopreservation of plant cells and organs. CRC Press, Boca Raton, FL.

Villiers, T.A. 1972. Ageing and the longevity of seeds in field conditions. p. 265-288. *In* W. Heydecker (ed.) Seed ecology. Butterworths, London.

Winters, H.F., and F. Rodriquez-Colon. 1953. Storage of mangosteen seeds. Proc. Am. Soc. Hortic. Sci. 61:304-306.

3 Regulatory Roles for Desiccation and Abscisic Acid in Seed Development: A Comparison of the Evidence from Whole Seeds and Isolated Embryos[1]

Allison R. Kermode, Maurice Y. Oishi, and J. Derek Bewley

University of Guelph
Guelph, Ontario

Embryo development can be divided conveniently into three confluent stages. Initially, during histodifferentiation, the single-celled zygote undergoes extensive mitotic division, and the resultant cells differentiate to form the basic body plan of the embryo. Thereafter, maturation occurs largely in the absence of further cell divisions, and is characterized by cell expansion and deposition of reserves (normally proteins, with lipid or carbohydrate) in the storage tissues. Maturation is terminated by drying, which results in a gradual reduction in metabolism as water is lost from the seed tissues and the embryo passes into a metabolically inactive, or quiescent, state. It is now evident that these events occur under the controlling influence and protection of the maternal environment. Interactions occur between the embryo and the surrounding seed tissues (i.e., the seed environment), which are in intimate contact with the maternal sporophyte (i.e. the maternal environment). These modulate and control the course of embryogeny. The extent to which interactions between the embryo and its immediate environment regulate development remains to be elucidated as are the signals that form the basis of these interactions. Knowledge is particularly lacking about the early stages of embryogenesis, when the initial differentiation/morphogenesis of the embryo is triggered. Somatic embryogenesis, the elaboration of a basic embryo plan from certain vegetative tissues, occurs in artificially created environments which are different from that present within the seed (Raghavan, 1986). However, such adventive embryos commonly fail to complete their developmental program and germinate without developing full-size, storage-laden cotyledons. Even zygotic embryos removed from the developing seed do not complete

---

[1] Contribution from Dep. of Botany, Univ. of Guelph, Guelph, Ontario, Canada N1G 2W1

Copyright © 1989 Crop Science Society of America, 677 S. Segoe Rd., Madison, WI 53711, USA. *Seed Moisture*, CSSA Special Publication no. 14.

their developmental program in culture, but instead germinate precociously and often fail to establish viable seedlings. Hence, an important role of the seed environment may be to maintain embryos in an developmental mode until they are fully formed and have accumulated sufficient reserves to permit successful germination and subsequent seedling establishment.

The mechanism whereby the seed environment retains the embryo in a developmental mode and suppresses germination is not known, although two potential regulatory factors have been studied, viz., abscisic acid (ABA) and restricted water uptake. Embryo culture has been used as a means of studying how these factors regulate development. When embryos are placed in culture they are separated from the endosperm or maternal factors that can be replaced artifically in the culture medium; hence it has been anticipated that a comparison of cultured embryos with those maturing in the seed allows a study of the relative contributions of environmental, maternal, and internal controls to development (Long et al., 1981). Following the completion of maturation the seed is released from these regulatory constraints, by which time it must have developed inherent control mechanisms that allow it to determine its own destiny, i.e., for germination and subsequent seedling establishment to occur after release from the maternal environment. Many questions remain to be answered concerning the control processes involved in embryogeny and germination. Is ABA or water stress an important factor involved in maintaining embryos in a developmental state? When does the developing seed establish itself as a potentially independent entity? (That is can it be capable of germination, yet be held in a developmental mode?) How are the influences of the maternal plant modulated or terminated at maturity? What initiates germination events and prevents the continuation of developmental processes in the mature dispersed progeny?

In this chapter we will advance our view that, for many seeds, the acquisition of desiccation-tolerance is an important event in establishing the potential autonomy of the seed, and that maturation drying is a critical event in terminating development, and overcoming those constraints by the maternal environment that maintain seeds in a developmental mode. The result is a "switch" in the seed's activities that become exclusively germination/growth related. Release of embryos from the influence of the maternal environment can be achieved without drying, in some cases naturally (even in the whole seed), and in others by removal of the embryo from its immediate environment. The developmental mode of the isolated embryo can be re-imposed in vitro by manipulations of the culture conditions. To what extent these provide insight into the controlling events in the developing seed in its natural environment will be outlined.

DESICCATION AND THE SWITCH FROM DEVELOPMENT TO GERMINATION

Desiccation as the Terminal Phase of Development

Seed development and germination are distinct physiological stages of the plant life cycle in which key metabolic events related to the status of stored reserves contrast markedly. As outlined earlier, development is characterized by the rapid accumulation of storage reserves, namely proteins (Fig. 3-3), and lipids or carbohydrates. During germinative growth, on the other hand, catabolic activities serve to mobilize these stored reserves. The products of these degradative processes are utilized both as the substrate and energy source for the growing seedling.

These distinct metabolic events implicate the involvement of a switch that effects the transition from development to germination. Several possibilities have been put forward to explain the nature of this switch. First, the transition in metabolism may be a preprogrammed temporal event that is genetically determined. Therefore, all events involved in development must occur before germinative events may proceed. Second, alterations in the levels of specific signals supplied by the maternal environment may regulate the shift from development to germinative events. In a related manner, the interactions of various seed tissues via chemical messengers may determine the activity of adjacent tissues. Finally, maturation drying, the net loss of water from the cytoplasm, an event that terminates normal seed development, may be causal in the cessation of developmental events and predispose the seed to germination upon rehydration (reviewed in Kermode & Bewley, 1986a). We will discuss these possibilities in relation to development and the transition from development to germination.

The suggestion that drying plays a critical role in this transition is based on a variety of observations. "Orthodox" seeds (see Chapter 6 in this book) undergo a period in a dry state that is interpolated between development and germination; the metabolic events associated with this latter event are activated by rehydration. Exceptions exist, such as the "recalcitrant" seeds (see Chapter 2 in this book), viviparous caryopses of mutant cereals, and certain weed seeds that can germinate without a requirement for drying. Normally, however, there is a temporal correlation between water loss during development and the onset of germinability of freshly harvested seeds. For example, seeds of french bean (*Phaseolus vulgaris* L.) and castor bean (*Ricinus communis* L.), when removed in the fresh state from the pod, will germinate only after achievement of maximum dry weight, which is just subsequent to the beginning of water loss in situ (Kermode & Bewley, 1985a; Dasgupta et al., 1982). Full germination is not achieved until later, when the seeds of these species have nearly completed the natural desiccation phase of their development (Kermode et al., 1986).

The use of mature dry seeds in experiments to determine the role of drying in the metabolic transition between development and germination has certain limitations. Since these seeds have already completed development, the factors or events that are likely involved in development (i.e., desiccation, signals from the maternal tissue, and genetic programming) may all have had an effect in determining the course of events upon imbibition. Hence, researchers have sought alternative approaches to study this problem. One such approach is to prematurely impose drying on the developing seed by removing it from the mother plant and allowing it to dry under controlled conditions. If drying elicits the switch from a developmental to a germinative metabolism, the prematurely dried seeds should respond similarly to mature dry seeds when imbibed, assuming they are tolerant of desiccation at earlier developmental times.

Whole Seeds: Requirement for Desiccation

Duchartre (1852) first documented the beneficial effects on germination of a drying treatment applied to developing cereal seeds. Subsequent work established that immature seeds of barley (*Hordeum vulgare* L.) and wheat (*Triticum aestivum* L.) will germinate if dried on the straw, while their fresh counterparts (i.e., those of the same developmental age, but not dried) will not (Harlan & Pope, 1922; King, 1976; Wellington, 1956). The beneficial effects of the drying treatment on germination, however, depend upon the age of the seeds at harvest time, and the rate of the imposed drying treatment. Seeds at all developmental stages are not equally tolerant of a prematurely applied drying treatment. Instead, they appear to acquire desiccation tolerance at roughly half-way through their development, e.g., in castor and french beans (Kermode & Bewley, 1985a; Dasgupta et al., 1982). At this stage, fresh seeds are unable to germinate. Seeds harvested during the first half of development and allowed to dry are unable to germinate and eventually deteriorate—they are desiccation intolerant. Ultrastructural studies indicate that the inability of prematurely dried french bean axes to germinate at the intolerant developmental stage is associated with rapid, irreversible damage to cellular membranes during drying, including mitochondrial and nuclear membranes (Dasgupta et al., 1982). Correspondingly, desiccation-intolerant seeds fail to recover full metabolic activity upon imbibition. Polysome recruitment and protein synthesis are both diminished. Similar damage is not apparent in rehydrated axes of desiccation-tolerant seeds.

The desiccation rate is also critical in determining the seed's ability to survive the drying treatment. Seeds of legumes (Ellis et al., 1987; Misra & Bewley, 1985; Adams et al., 1983); maize (*Zea mays* L.) (Oishi & Bewley, 1987, unpublished data; and castor bean (Kermode & Bewley, 1985a) are unable to withstand rapidly imposed drying (over silica gel or under regimes similar to ambient laboratory conditions) at early developmental stages. For instance, when maize kernels and castor bean seeds are rapidly dried during most of the desiccation-tolerant developmental stage, before natural maturation drying takes place, they show no germinability. This contrasts with seeds at the same developmental stage dried slowly over saturated salt solutions or attached to the cob or pod, where full germinability is evident. Tolerance of rapid drying occurs only at or near the completion of reserve deposition as indicated by attainment of maximum dry weight (Kermode & Bewley, 1985a; Misra & Bewley, 1985; Ellis et al., 1987). Although the underlying reasons for the differing responses to rapid and slow drying are not understood, the former appears to increase rates of solute leakage through membranes upon imbibition (Adams et al., 1983), suggesting that damage to cellular membranes may be involved (see Chapter 1 in this book).

Desiccation Elicits Changes in Protein Synthesis

It is apparent that prematurely imposed drying enables the seed to assume a germinative response upon imbibition. The question may be posed as to whether imbibition leads to the direct initiation of germinative events found in mature seeds or whether it leads to the resumption and possible completion of reserve accumulation, including storage synthesis, before the onset of germination. It has been established that developing axes of french bean synthesize a number of unique proteins, including the 7S storage protein phaseolin, whereas mature germinating axes synthesize proteins that are not evident during development. At 32 d after pollination (DAP), developing axes are capable of germination if prematurely dried before imbibition. This change from developmental to germinative events is mirrored in the pattern of proteins synthesized by the germinating seeds (Dasgupta & Bewley, 1982). As early as 12 h after imbibition, the axes of such prematurely dried seeds synthesize proteins characteristic of germinating axes from mature seeds imbibed for the same period; development-associated proteins, including phaseolin, are not synthesized. Hence, it is evident that drying plays a role in the permanent suppression of developmental protein synthesis and in inducing germinative protein synthesis. Moreover, similarly treated 22 DAP axes (which are not desiccation tolerant) fail to complete this switch in direction of metabolism; instead, they synthesize proteins characteristic of both germination and development, albeit to considerably reduced extent. Thus, the acquisition of competence of the genome to fully respond to drying is acquired simultaneously with the development of desiccation tolerance.

The castor bean endosperm (a nongerminating seed structure) also exhibits a transition in protein synthesis in response to premature drying (Kermode & Bewley, 1985b; Kermode et al., 1985). Endosperms of mature seeds imbibed for up to 5 h synthesize soluble proteins typical of those produced during development. Thereafter, the synthetic pattern changes and proteins identified as being unique to germination and growth are made (Fig. 3-1A). In a similar manner, synthesis of those insoluble proteins that are characteristic of late-development resumes upon imbibition and continues longer into germination and growth than the synthesis of developmental soluble proteins—even as long as 48 to 72 h after imbibition (Fig. 3-1A).

When endosperms from prematurely dried 30 to 40 DAP seeds are rehydrated, the pattern of synthesis of the soluble and insoluble proteins is virtually identical to that exhibited by the endosperms of imbibed mature seed (compare Fig. 3-1A and 3-1B). Furthermore, during the first 48 to 72 h of rehydration, the 30 and 40 DAP endosperms synthesize insoluble proteins that are characteristic of late development (45-50 DAP), even though they were desiccated during development prior to the onset of the synthesis of these proteins. Thus, premature drying of seeds in the desiccation-tolerant developmental stage leads to the same asynchronous changes in soluble and insoluble protein synthesis as upon imbibition of the mature seed.

It is noteworthy that developing seeds do not need to complete reserve accumulation before they are able to be switched by drying into a germinative

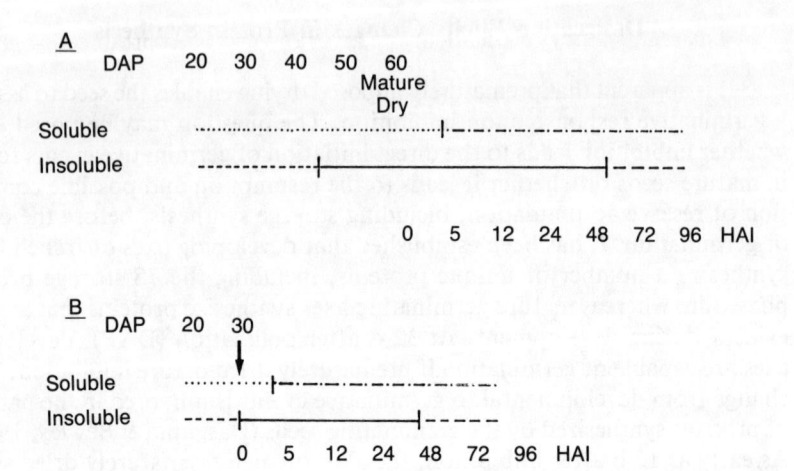

Fig. 3-1. Temporal changes in protein synthesis in the insoluble and soluble protein fractions of castor bean endosperm during (A) normal development and germination/growth, and (B) development, terminated by premature desiccation, and subsequent germination/growth. The discontinuities in the lines represent changes in the patterns of protein synthesis. Similarly denoted lines represent similar patterns of protein synthesis over the time periods indicated. For example, insoluble protein synthesis during normal development is intense and yields largely crystalloid storage proteins until 45 DAP (---). Thereafter, until 60 DAP and subsequently until approximately 60 HAI, a low amount of the crystalloid storage protein and some late developmental insoluble proteins are synthesized (—). A low level of synthesis of insoluble proteins unique to growth is evident after 60 HAI (--).

DAP = Days after pollination. HAI = Hours after imbibition. Arrow (B) marks time of prematurely imposed desiccation. Based on data published in Kermode and Bewley (1985b) and Kermode et al. (1985).

mode. At 25 DAP, castor bean seeds are tolerant of desiccation and yet have accumulated only 17% of their total protein reserves and 13 and 3% of their final lipid and phytin content, respectively (Greenwood et al., 1984). Protein synthesis does not substantially add to the reserves of the seed upon rehydration, and thus it is evident that the completion of developmental events is not a prerequisite for germination. Hence, the switch to germination is not a preprogrammed event that relies upon the completion of all developmental events.

In addition to the changes in the protein synthetic pattern, premature drying also appears to predispose the seed to the mobilization of reserves of the storage tissues upon imbibition. Catabolism of storage proteins in the castor bean endosperm and cotyledons is associated with an increased activity of enzymes responsible for their mobilization including LeuNAase, an aminopeptidase (Kermode & Bewley, 1985b, 1986b). In the oil-storing seeds of soybean [*Glycine max* (L.) Merr.] and castor bean, increased isocitrate lyase and malate synthetase activity also have been identified in mature germinated seeds and in those rehydrated from the prematurely dried state (Adams et al., 1983; Kermode & Bewley, 1985b, 1986b). Enzyme activities do not increase in prematurely harvested seeds that are maintained in the fully hydrated state.

The competency of the aleurone layer of cereals to produce α-amylase in response to gibberellin (GA) is induced by both maturation drying and prematurely imposed drying (Evans et al., 1975; Armstrong et al., 1982). Sensitivity to GA is attained in developing tissues of wheat if the seed is first desiccated to a water content of <30%. Drying may effect either a change in membrane lipid and protein conformation, thus increasing sensitivity to GA (Norman et al., 1982), or a decrease in the level of a growth regulator that is antagonistic to GA (Cornford et al., 1986).

It has been argued that the switch to germination and growth is not directly due to drying but to a period of detachment from the mother plant. However, immature castor bean seeds removed from the plant and maintained in the hydrated state on water for a period of time equivalent to that used to achieve drying, do not germinate (Kermode & Bewley, 1985a). Furthermore, isolated seeds maintained in this fully hydrated state fail to exhibit the metabolic changes that normally follow germination, namely increased LeuNAase or isocitrate lyase activity in castor bean (Kermode & Bewley, 1985b, 1986b) and α-amylase activity in wheat (Armstrong et al., 1982). Such seeds differ from prematurely dried seeds only in terms of their hydrational state after detachment. Therefore, isolation of seeds from the maternal environment alone is not sufficient to induce germination-related events.

Desiccation Induces Changes in Translatable mRNAs

The amount and composition of the cell's mRNA complement reflects changes in translational and transcriptional control during development and germination. Changes in the extent of storage protein synthesis during seed development appear to be related to the amount of appropriate mRNAs available (Goldberg et al., 1981a, 1981b) that is indicative of regulatory processes effected primarily at the level of transcription (i.e., mRNA synthesis) and mRNA stability. In the previous section, we outlined that changes in protein synthesis are indicative of a switch in genomic activity (i.e., a permanent suppression of developmental protein syntheses and an induction of germination/growth-related synthesis) brought about by premature desiccation. Hence, it is pertinent to examine the molecular level of control at which desiccation effects this switch in metabolic activity. It could be effected primarily at a transcriptional (mRNA synthesis) level, i.e., premature drying could result in a loss of mRNAs for developmentally related proteins (as well as a permanent suppression of their production) and an induction of those mRNAs for germination proteins. Or, alternatively, the switch could be controlled at the level of translation, (i.e., the mRNAs for both development and germination are present within the cells of the germinating seed) with only the latter being selectively translated. To examine these possibilities, the effects of premature desiccation and rehydration upon the fate and stability of the translatable mRNA fraction of the castor bean endosperm have been analyzed (Kermode, 1988, unpublished data).

During normal seed development, there are quantitative and qualitative changes in the mRNA complement of the endosperm. In particular, messages coding for high MW 50 to 60 kDa proteins (possible storage protein precursors) decline substantially in amount during maturation drying. A similar decline occurs during premature desiccation of immature seeds. Following rehydration of these prematurely dried seeds, the pattern of proteins synthesized in vitro by the extracted mRNAs is identical to that carried out by messages from mature seed endosperms during and following germination. Some of the messages associated with development that persist in the endosperm following natural or imposed desiccation decline during early imbibition. New mRNAs can be detected following the completion of germination and during early seedling growth, including the production of those messages coding for proteins presumed to be involved in reserve mobilization within the endosperm. In a similar vein, in aleurone layer cells from immature wheat seed, premature desiccation and rehydration induces the accumulation of mRNA coding for the post-germinative enzyme α-amylase in response to GA (Cornford et al., 1986). Interestingly, however, this induction of GA responsiveness in immature aleurone layer cells does not extend to the production of other enzymes associated with normal post- germinative growth, including acid phosphatase, protease, and ribonuclease (Cornford et al., 1986). This could be the consequence of a differential sensitivity of the genome to premature desiccation such that the messages for some enzymes are induced and those for others are not. Presumably, as the seed matures, the sensitivity of the genome to drying changes in relation to these enzymes since they are produced in the aleurone layers of mature germinated seeds.

A switch in message population is elicited also by premature desiccation and rehydration of french bean axes (at a tolerant developmental stage). For example, the mRNAs for several major developmental proteins (including the storage protein phaseolin) are degraded during early rehydration following premature drying. New mRNAs, coding for germination proteins, are available for utilization in these rehydrated axes; messages that were not present either before drying or in the dry state (Misra & Bewley, 1985).

These studies are indicative of a control by drying primarily at the level of translatable mRNAs (Fig. 3-2). Natural or imposed desiccation may cause a decline in, or loss of some developmental messages due to their reduced stability and/or increased degradation, and a decline in their synthesis. Those messages coding for storage proteins that persist in the prematurely dried (or mature dry) seed appear to be labile (or unprotected) and decline rapidly during early germination stages. At this time, these messages cannot be replaced because their transcription has been switched off directly or indirectly due to the drying treatment. The appearance of new mRNAs, that are for germination- and growth-related proteins, is suggestive of an induction of transcription of their genes that is effected upon rehydration, following premature desiccation. It is noteworthy that in both normally and prematurely dried seeds of castor and french beans, the same temporal sequence of changes to the mRNA population occurs during subsequent germination. Thus,

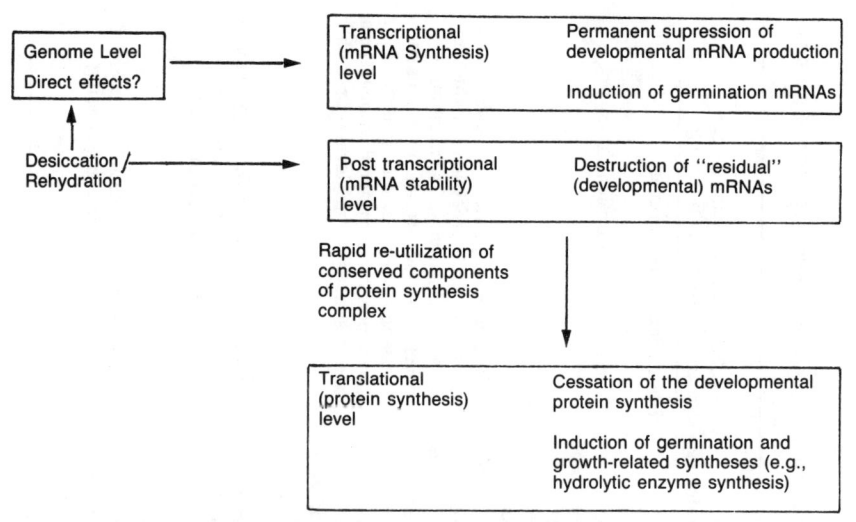

ROLES FOR DESICCATION AND ABA IN SEED DEVELOPMENT

```
┌──────────────────┐      ┌──────────────────┐   ┌──────────────────────────────┐
│ Genome Level     │ ───► │ Transcriptional  │   │ Permanent supression of      │
│ Direct effects?  │      │ (mRNA Synthesis) │   │ developmental mRNA production│
└──────────────────┘      │ level            │   │                              │
         ▲                │                  │   │ Induction of germination mRNAs│
         │                └──────────────────┘   └──────────────────────────────┘
 Desiccation /────────►   ┌──────────────────┐   ┌──────────────────────────────┐
 Rehydration              │ Post transcriptional│ Destruction of "residual"    │
                          │ (mRNA stability) │   │ (developmental) mRNAs        │
                          │ level            │   │                              │
                          └──────────────────┘   └──────────────────────────────┘
  Rapid re-utilization of
  conserved components
  of protein synthesis
  complex
                          ┌──────────────────┐   ┌──────────────────────────────┐
                          │ Translational    │   │ Cessation of the developmental│
                          │ (protein synthesis)│ │ protein synthesis            │
                          │ level            │   │                              │
                          │                  │   │ Induction of germination and │
                          │                  │   │ growth-related syntheses (e.g.,│
                          │                  │   │ hydrolytic enzyme synthesis) │
                          └──────────────────┘   └──────────────────────────────┘
```

Fig. 3-2. A summary of the effects of desiccation in eliciting a switch from developmental protein synthesis to germinative protein synthesis: potential points of control.

premature drying of the developing seed during the tolerant stage does not appear to adversely affect expression of the genome.

We conclude that all of the changes in the message population, from those unique to development to those specific to germination, do not occur during drying per se. Rather, the appearance of messages for germination appear after hydration of the dry seed (after a lag period of a few to several hours), and the loss of developmental-specific messages is completed at this time. Thus, rehydration following premature desiccation appears to be the crucial event leading to the loss of any long-lived developmental messages—i.e., those that remain stable throughout desiccation. However, it is likely that the cessation of developmental mRNA synthesis takes place earlier—i.e., during drying itself.

In concluding this section, a summary of seed responses to desiccation is presented in Table 3-1. Figure 3-2 summarizes how desiccation affects a switch in protein synthesis. A more complete account of these topics is found in Kermode and Bewley (1986a).

EMBRYO RESPONSES TO ISOLATION DURING DEVELOPMENT

Germination in the Absence of a Drying Period

Imposed desiccation (with few exceptions, as previously noted) is required for the cessation of development and the promotion of germination and growth of the immature whole seed. There are conditions, however, where water loss is not a prerequisite for the initiation of germination/growth events. For example, immature embryos of rape (*Brassica napus* L.), french bean, cotton (*Gossypium hirsutum* L.), soybean, rice (*Oryza sativa* L.), and wheat

Table 3-1. Summary of responses to desiccation, imposed prematurely, or naturally during maturation drying, on the metabolism and germination/growth potential of seeds.

Development stage	Water status	Tolerance of desiccation during development	Known cellular responses to premature or maturation drying	Possible mode of action of premature or maturation drying
Histodifferentiation	Water content and fresh wt. increase	Intolerant	Loss of cellular integrity	Disruption of membranes
Cell expansion	Increases during initial phase; declines thereafter due to reserve accumulation	Achieved about the mid-point of development	When tolerant, premature germination occurs. Release from inhibition of germinative and postgerminative processes. Responses as to maturation drying	As in normal maturation drying
Maturation drying	Rapid decline in fresh wt. and water content	Tolerant	Decline in developmental protein synthesis, and in developmental mRNAs. Termination of developmental events, stimulus for germination (Fig. 3-3)	Decline in ABA production and/or flow, enhanced ABA breakdown, reduction in ABA sensitivity. Termination of developmental, (storage protein) synthesis via effects at the transcriptional posttranscriptional, translational and posttranslational level (Fig. 3-2)
Mature seed	Low water content	Tolerant	Quiescent. Conservation of components of protein-synthesizing complex	
Germination	Three-phase imbibition Increase in water content		Utilization of conserved components of protein synthesizing complex Developmental mRNAs destroyed as germinative ones synthesized	Promotion of transcription of germinative mRNAs. Suppression of transcription of developmental mRNAs
Postgerminative growth	Increase in water content		Promotion of postgerminative enzyme synthesis. Increased sensitivity to GA (e.g., aleurone layers)	Compositional/conformational changes in membranes, enhancing binding site availability

will germinate when excised from the developing seed and placed in tissue (embryo) culture conditions (Ihle & Dure, 1972; Quebedeaux et al., 1976; Crouch & Sussex, 1981; Long et al., 1981; Triplett & Quatrano, 1982; Ackerson, 1984a; Stinissen et al., 1984). The isolated embryos may germinate immediately, e.g., cotton (Ihle & Dure, 1972); or continue to develop and accumulate storage proteins for several days in culture, as they would in the intact seed, before becoming briefly quiescent and then germinating (e.g., french bean) (Long et al., 1981). This capacity for precocious germination of the (nondried) isolated embryo is generally acquired after histodifferentiation and organ/meristem formation have been completed, about a third to a half of the way through seed development, roughly coincident with the acquisition of desiccation-tolerance. Thus, the embryo acquires a competence to respond to "germination cues" well before the completion of major developmental events (such as reserve accumulation and the onset of normal maturation drying). Presumably, the environment of the developing embryo within the seed is normally such that it encourages development and prevents germination; removal of the embryo from this environment leads to the commencement of germination and growth processes. Premature drying can also overcome the inhibitory influence of the seed environment; however, sensitivity to this cue must first be acquired.

Precocious germination induced by embryo isolation does not always lead to a permanent supression of developmental metabolism as occurs in normal (i.e., mature seed) germination. For example, embryos of rape (isolated in the fresh state from the seed and placed in tissue-culture conditions) do not acquire the ability to grow into normal seedlings unless they are isolated near maturity, i.e. 50 d postanthesis (DPA) or later. Precocious germination of less mature (30–40 DPA) embryos leads to an abnormal growth pattern and the continuation of storage protein accumulation even in newly formed tissues (Finkelstein & Crouch, 1984). Hence, both developmental and germination/growth events occur simultaneously (unlike in mature tissue) (Finkelstein & Crouch, 1984). The response of immature embryos to their isolation (and extent of overlap between events normally exclusive to the developmental-or the germination/growth program) appears to vary between species and often with the age at which embryos are isolated from the seed, or the culture conditions under which they are maintained. This is illustrated in Table 3-2 for embryos of rape and castor bean whose response to isolation changes with age at excision during development, and for cotton embryos, in which the culture conditions elicit a different response. It is not known to what extent this variation in behavior reflects underlying differences in the degree to which the cessation of developmental events is under internal control of the embryo (as opposed to external factors provided by the mother plant and/or surrounding seed tissue).

The concurrent expression of developmental-and germination/growth-related genes in some isolated embryos in culture contradicts the idea that the two programs are mutually exclusive, with a simple switch controlling the entire suite of characteristics of each program. It also indicates that events required for the cessation of development may be distinct from those involved

Table 3-2. Four categories of embryo responses to isolation during development.

	Continuation of developmental processes; no switch to germination/growth-related events	Period of continued developmental processes; a switch to germination/growth-related events after a lag	Cessation of developmental processes; prompt switch to germination/growth-related events	Simultaneous developmental and germination/growth-related syntheses
Species	Rape (*Brassica napus* L.) embryos, <30 DPA[†]	Rape hypocotyls of 30-35 DPA embryos	Rape embryos >50 DPA (nearly mature)	Rape embryos, 30-40 DPA
Medium[‡]	Basal	Basal	Basal	Basal
Morphological response	Secondary embryo production	Hypocotyl growth after 3 wk	Normal germination and growth	Germination. Secondary cotyledon formation, instead of leaves.
Storage proteins	Low level of synthesis	Accumulation until hypocotyl growth commences, then degradation	Rapid degradation as in normal postgerminative growth	Accumulation even in newly formed roots and cotyledon tissues. No degradation.
Reference	Finkelstein & Crouch, 1984	Finkelstein & Crouch, 1984	Finkelstein & Crouch, 1984	Finkelstein & Crouch, 1984
Species	Castor bean (*Ricinus communis* L.) embryos, <30 DAP[§]	Castor bean embryos, 30 DAP	Castor bean embryos, >35 DAP	
Morphological response	No germination	Germination and growth after 4-7 d	Germination and growth commence immediately	
Storage proteins	—	—	Degradation	
Reference	Kermode & Bewley, 1987a	Kermode & Bewley, 1987a	Kermode & Bewley, 1987a	
Species			Cotton (*Gossypium hirsutum*) embryos, (mid-maturation)	Cotton embryos (mid-maturation)
Medium			Water	Enriched medium
Morphological response			Precocious germination	Precocious germination
Storage proteins			Synthesis ceases. Germination/growth-related proteins appear	No synthesis of storage proteins. Synthesis of proteins characteristic of late embryo development and postgerminative growth
Reference			Ihle & Dure, 1972; Dure et al, 1981	Choinski et al., 1981; Galau et al., 1986

[†] Days post-anthesis. [‡] Medium on which embryos were placed after isolation. [§] Days after pollination.

in the initiation of germination and growth (M. Crouch, 1987, personal communication). Thus while premature desiccation and isolation elicit the same morphological response (viz., germination), the metabolic responses of these two treatments appear to be different. Desiccation and subsequent rehydration (of immature embryos within whole seeds) results in a permanent suppression of developmental metabolism and an induction of that which is germinative in nature; this is not necessarily the case in immature embryos that germinate upon isolation. It is pertinent to ask whether the isolated embryo fails to switch completely in its metabolism because it has not been subjected to desiccation. Or, does the embryo normally require signals from surrounding seed tissues (that may also need to be dried) to terminate developmental syntheses and irreversibly switch to a germination and growth program? Our preliminary studies with isolated castor bean embryos (Kermode & Bewley, 1987a) indicate that both drying and the presence of the surrounding seed tissues are important.

Role of the Seed Environment: Abscisic Acid and Osmotic Potential (ψ_s)

From the observed tendency of immature embryos to germinate in culture, it has been suggested that the seed environment (rather than the embryo itself), enforces embryo maturation and suppresses germination. The controlling factor(s) is not known, although addition of the plant hormone, abscisic acid (ABA) at physiological concentrations to developing embryos of several species (in culture) prevents their precocious germination and encourages the continuation of developmental processes (Crouch & Sussex, 1981; Eisenberg & Mascarenhas, 1985; Long et al., 1981; Triplett & Quatrano, 1982; also reviewed in King, 1982; Quatrano, 1986; Walton, 1980/81). In several cases (but by no means all), the addition of ABA to the medium is a prerequisite for normal embryogeny to proceed (Ammirato, 1977). In such cultures, ABA inhibited precocious germination but enhanced the following:

1. The accumulation of storage reserves, particularly storage proteins, e.g., in french bean, rape, soybean, and wheat (Ackerson, 1984a; Bray & Beachy, 1985; Crouch & Sussex, 1981; Eisenberg & Mascarenhas, 1985; Finkelstein et al., 1985; Triplett & Quatrano, 1982; Williamson et al., 1985).
2. Development of enzyme activities in late embryogenesis (e.g., malate synthetase in cotton) (Choinski & Trelease, 1978; Choinski et al., 1981).
3. Production of certain mRNA and protein fractions characteristic of late embryogenesis in cotton (Dure et al., 1980; Galau & Dure, 1981; Galau et al., 1986).

These late proteins are not synthesized when embryos precociously germinate (on water) in the absence of ABA (Galau & Dure, 1981). It has been surmised that ABA (synthesized within the surrounding seed tissues or supplied by the mother plant through the vascular supply in the seed coat) maintains embryo metabolism in a developmental (i.e., largely anabolic) mode (Fig. 3-3).

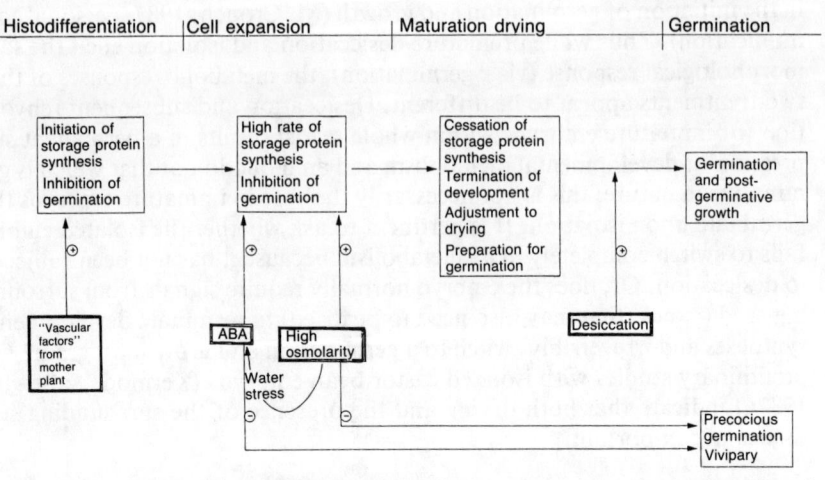

Fig. 3-3. Events during the development and germination/growth of seeds that are affected by desiccation, high osmolarity, or ABA. For explanation refer to Tables 3-1 and 3-3, and in the text. Plus signifies where factors promote events at the point of development, and minus signifies where processes are inhibited by the factors.

One component of the chemical environment surrounding the developing embryo is the osmotic environment. This environment appears to be highly specialized. It is known, for example, that the liquid endosperm in which developing embryos are constantly bathed has generally quite negative water and osmotic potentials (Raghaven, 1986; Yeung & Brown, 1982). Precocious germination is often suppressed if a culture medium has a high osmotic potential (presumed by some to act by altering endogenous ABA levels), and here again an embryonic morphology is maintained (Crouch & Sussex, 1981; Obendorf & Wettlaufer, 1984). Thus, another regulatory control over embryo development may be restricted water uptake caused by the high osmolarity of surrounding seed tissues (Fig. 3-3).

A prompt switch to germination and growth of excised cultured embryos is not observed in all species (response B Table 3-2). French bean axes in culture also continue to develop embryonically for a period of time under their own intrinsic control (response B) (Long et al., 1981). If ABA or water stress is indeed responsible for preventing precocious germination, it may be that a high level of ABA is maintained or synthesized internally by embryonic axes of french bean, while in other embryos the maternal environment supplies ABA and/or causes water stress (Long et al., 1981).

The transition from development to germination, as effected by external or internal agents has been monitored in several ways, including studies of metabolism and physiological/morphogenetic changes. These are considered below under convenient subheadings. Reference is made to Fig. 3-3, and accompanying Table 3-3, that point particularly to where ABA may modulate seed-developmental events.

Table 3-3. The levels of, and sensitivity to, ABA of seeds during development and germination/growth, and known responses to application of this hormone. Refer to Fig. 3-3 for control points where ABA promotes developmental events (plus), and inhibits the switch from development to germination (minus).

Developmental stage	ABA levels	ABA sensitivity	Known responses to ABA	Possible mode of action of ABA
Mid-cell expansion	High	Maximum	High rate of storage protein synthesis. Prevention of precocious germination	Modulation of storage protein mRNAs
Prior to onset of maturation drying (physiological maturity)	High	Maximum	Prevention of vivipary and hydrolysis of storage reserves	Prevention of water uptake into embryo
Maturation drying	Precipitous decline in most seeds	Decline in some seeds, not in others	Usually ineffective in preventing decline in synthesis of storage proteins and their mRNAs	Genes for storage proteins insensitive to ABA
Mature dry seed	Very low (in most nondormant seeds)			
Germination	Very low (in most nondormant seeds)	High in seeds whose germination is inhibited by exogenous ABA	Germination inhibited in some species. Developmental protein synthesis. Inhibition of cell elongation Germination not inhibited in other species. None, due to lack of receptor sites, or ABA-sensitive components	Protein synthesis integral to germination inhibited
Postgerminative growth	Very low, unless water stress imposed	Loss of sensitivity to exogenous ABA with time in some species, but not others	Inhibition of reserve mobilization	Inhibition of hydrolytic enzyme synthesis

Reserve Deposition

A large component of growth or maturation of the developing embryo is associated with the laying down of reserves; a role for ABA in this respect has been suggested, particularly in promoting the synthesis of storage proteins (Fig. 3-3, Table 3-3, cell expansion phase). Evidence for this comes mainly from in vitro studies. For examples, in french bean, Sussex and Dale (1979) culturing whole embryos found an increase in storage protein (vicilin) synthesis and accumulation when ABA was included in the medium. Crouch et al. (1985) and Finkelstein et al. (1985) found that ABA (or high concentrations of osmotica) allowed embryonic development of rape to proceed in the absence of the endosperm and maternal plant including continued expression of the embryo-specific storage protein genes, napin and cruciferin. However, the ability of ABA to maintain storage protein gene expression was dependent upon embryo age; a loss in sensitivity occurred in embryos at later maturation stages (Finkelstein et al., 1985). Immature embryos of wheat are prevented from germinating precociously when ABA is supplied in the culture medium; its presence also appears to enhance the synthesis of the embryo-specific lectin, wheatgerm agglutinin (Triplett & Quatrano, 1982; Raikhel & Quatrano, 1986) and E_m protein (an albumin storage protein abundant in the mature embryo) (Williamson et al., 1985). The effects of ABA on the synthesis and accumulation of the 11S storage proteins (and their mRNAs) in cultured soybean embryos are stage-dependent (Eisenberg & Mascarenhas, 1985). Abscisic acid is required for high levels of storage protein gene expression in early and mid-maturation embryos. However, at younger developmental stages (e.g., cotyledon stage), ABA triggers a decline in 11S protein mRNA; it is ineffective in reversing the decline in mRNA levels in embryos cultured at the late maturation stage. Cultured soybean cotyledons (detached from the embryonic axis) specifically increase the accumulation of the β-subunit of the 7S storage protein, β-conglycinin (and its mRNA) in response to exogenous ABA; accumulation of the α- or α-' subunits remains unaffected (Bray & Beachy, 1985).

The molecular levels at which ABA maintains or enhances storage protein synthesis/accumulation in isolated embryos are various. There is evidence for regulation of transcription (Bray & Beachy, 1985; Eisenberg & Mascarenhas, 1985; Williamson et al., 1985) but control at the post-transcriptional, translational, and post-translational levels has also been suggested (Crouch et al., 1985; Finkelstein et al., 1985; Quatrano, 1986; Williamson et al., 1985).

Evidence for a comparable role of endogenous ABA during normal seed development (i.e., in situ on the mother plant) is mainly restricted to the finding that in some seeds (e.g., wheat, rape, soybean, and french bean) the highest ABA levels are present during the most active phase of seed enlargement and deposition of reserves (Finkelstein et al., 1985; Hein et al., 1984; Hsu, 1979; King, 1976; Quebedeaux et al., 1976; Radley, 1979). This characteristic is not universal, however (Dure et al., 1980; Dure & Galau, 1981). In isolated immature cotton embryos treated with ABA, some embryogenic protein synthesis associated with late development occurs (Galau et al., 1986), but this does not include the major storage proteins. Furthermore, the major storage proteins are not synthesized at high levels during the embryogenesic stage (in situ) when the ABA concentration is high (Dure et al., 1980; Dure & Galau, 1981). Other factors, therefore, must be involved in regulating the expression of storage protein genes in cotton (Dure & Galau, 1981). Davies and Bedford (1982) have studied the effect of ABA on protein synthesis in cultured embryos of pea (*Pisum sativum* L.) and found that ABA had no effect on the accumulation of the storage protein legumin. However, synthesis of vicilin storage protein in pea is enhanced by ABA applied to seeds in situ (Schroeder, 1984).

As mentioned earlier, a high osmolarity can substitute for ABA in suppressing precocious germination and allowing embryo maturation to continue. In rape embryos, high osmoticum can simulate the seed environment over a wider range of developmental stages than ABA (Finkelstein & Crouch, 1986). Furthermore, high endogenous ABA levels are not required in osmotically treated embryos for inhibition of germination or synthesis of elevated levels of storage protein (cruciferin) and its mRNA. Thus, the ABA effects on developmental gene expression may be indirect and a mechanism involving control of water uptake by ABA has been invoked (Finkelstein & Crouch, 1986) (Table 3-3).

There are some obvious limitations to studies on the effects of exogenous ABA on cultivated embryos, and likewise to those that attempt to correlate peak hormone levels with metabolic events occurring in normally developing seeds on the mother plant. In general, such studies provide little information on the regulatory role of ABA in developing seeds in situ. Therefore, attempts have been made to test the relationship between ABA and developmental events by manipulating endogenous levels of this hormone via environmental, chemical, or genetic means (Fong et al., 1983; Groot, 1986; Karssen et al., 1983, Koornneef et al., 1984). For example, developing seeds or embryos can be made deficient in endogenous ABA by use of the chemical fluridone (an inhibitor of carotenoid biosynthesis) (Creelman & Zeevaart, 1981). The resulting change in phenotype is presumed to be due to interrupted ABA biosynthesis (rather than other processes), particularly if exogenous ABA reverses or counteracts the fluridone effects. For example, direct application of fluridone to cultured soybean cotyledons results in a decline in the β-subunit of β-conglycinin and its corresponding mRNA along with the endogeneous ABA level, an effect that is overcome by addition of ABA (Bray & Beachy, 1985). Similar inhibitory effects of fluridone on storage protein synthesis have been noted in cultured cotyledons of broad bean (*Vicia faba* L.) (Barratt, 1986).

Isogenic lines differing in endogenous ABA content (or ABA sensitivity) offer a more decisive means of elucidating its regulatory functions in situ. Studies with an ABA-deficient tomato (*Lycopersicon esculentum* L.) mutant show that despite strong reductions of ABA content in the developing seeds, neither fresh nor dry weight, nor the composition and content of reserve proteins are affected (Groot, 1986). Hence, the role of ABA in regulating storage protein synthesis in the developing seed on the mother plant, is still inconclusive and controversial.

Adventive embryos (that have never been exposed to the endosperm, seed coat, or maternal vascular supply) have been studied to determine the importance of external influences early in development on the pattern of storage protein synthesis in rape (Crouch, 1982). These embryos (e.g., those induced to form from microspores in anther culture) contain the major (1.7S and 12S) storage proteins. Hence, the capacity to express embryo-specific (storage protein) genes is possible in the total absence of any influence from the endosperm or maternal vascular supply. Although microsporic embryos of rape commence synthesis of storage proteins much earlier than zygotic embryos, these proteins are not accumulated to nearly as high a level. Thus, an important role of the seed environment may be to modulate the synthesis of embryo-specific products.

Prevention of Germination

From studies on the effects of ABA on embryos in culture, it appears that while normal embryogenic development can occur in its presence, in many cases germinative growth (i.e., the longitudinal extension of the axis) cannot (Bewley & Black, 1985). Therefore, ABA provided by the seed environment might be the factor that prevents the embryo from passing directly from embryogenesis to germination while still on the mother plant, without an intervening rest period (Sussex, 1975; Robichaud et al., 1980; Bewley & Black, 1985). Correlations between ABA content and germinability of the developing embryo are indicated in some species. Immature seeds in early developmental stages generally cannot germinate, but the ability to do so is acquired at ages closer to full maturity (Bewley & Black, 1985). Abscisic acid levels are higher in young, nongerminable seeds than in older ones, e.g., in pea, soybean, wheat, and french bean (Eeuwens & Schwabe, 1975; King, 1976; Quebedeaux et al., 1976; Radley, 1979; Van Onckelen et al., 1980). Ackerson (1984b) found that immature embryos of soybean can be induced to germinate by treatments that deplete the endogenous pool of ABA. However, a decrease in endogenous levels of ABA is not always correlated with germination (Braun & Khan, 1975).

In some cases, levels of endogenous ABA correlate with the capacity of embryos (at various developmental stages) to germinate in vitro (Ackerson 1984a; Finkelstein et al., 1985; Prevost & LePage-Degivry, 1985). In rape, there is a strong positive correlation between endogenous ABA levels at the time of embryo excision during development and the lag period prior to the start of axial extension in culture. However, ABA levels in these embryos (at any development stage) decrease dramatically (in a short time) following their excision and placement in culture (Finkelstein & Crouch, 1984). In cultured embryonic axes of soybean (Quebedeaux et al., 1976) and french bean (Long et al., 1981), elongation always begins when the axes are at about the same developmental age (regardless of their time of excision)—and this may not necessarily be related to the amount of ABA that is present, but to other changes (Black, 1983).

Continued axial growth of the embryo (germination) rarely takes place within the (maturing) seed on the plant. Two exceptions are (i) vivipary and (ii) the preharvest sprouting of some cultivars of wheat and barley (and several other species). Some of the most interesting and suggestive evidence for the role of ABA in the control of germination of the developing seed is provided by cases of vivipary. This condition is characterized by germination of the embryo within the fruit on the mother plant. There is an uninterrupted progression from embryogenesis to germination with little or no intervening cessation of growth (quiescence) and, in most cases, little or no dehydration (Fig. 3-3).

The embryos of viviparous mutants of maize, for example, either contain less ABA, or are less sensitive to applied ABA than the wild type (McDaniel et al., 1977; Neill et al., 1986, 1987; Robichaud et al., 1980; Smith et al., 1978). The *vp1* mutant of maize has normal carotenoid and ABA levels but requires higher levels of exogenous ABA to inhibit germination, indicating relative insensitivity to ABA (Robichand et al., 1980). Here, there is an almost complete independence of the embryo from the potentially inhibitory effects of the surrounding seed tissue (i.e., the endosperm). For example, it is the genotype of the embryo, independent of that of the endosperm, that determines the expression of vivipary and it is possible to obtain viviparous embryos in "dormant" (wild-type) endosperm and vice-versa (Neill et al., 1987). If this extends to embryos of nonviviparous seeds, it may be that the actual mechanism responsible for the prevention of germination resides within the embryo itself. The endosperm may provide an environment that encourages ABA synthesis within the embryo or provides some signal to which the embryo responds by continuing embryogenic metabolism and development (Neill et al., 1987).

It is interesting that when the endosperm is of the mutant type, *vp1* reduces the expression of several genes normally expressed in late aleurone layer development (Dooner, 1985). However, the connection, if any, between ABA sensitivity and the pleiotropic endosperm effects is unknown. Possibly *vp1* is a regulatory gene that induces the expression of many structural genes, including that for the ABA receptor. Alternatively, a number of different genes may be under the control of ABA, and in the absence of an appropriate hormone receptor their expression will be muted.

Vivipary can also be induced in wild-type individuals of maize by application of fluridone, which lowers ABA levels (Fong et al., 1983). Such a treatment has been used to determine the time in maize seed development when ABA is important for inhibiting precocious germination. Although carotenoid and ABA synthesis was always disrupted (when fluridone was applied to kernels at different times during development) only treatments between 9 and 13 d after anthesis resulted in subsequent vivipary (Fong et al., 1983). Relative insensitivity to ABA appears to be the case in mangroves (e.g., *Rhizophora mangle* L.) where vivipary is a normal occurrence. Unusually high concentrations of ABA are required to inhibit the growth of excised embryo-seedlings of this species (Sussex, 1975). Hence, a transition from development to germination during embryo maturation, in the absence of maturation drying, may be permitted by a low level of endogenous ABA (either within the embryo itself or in the surrounding seed tissues) or to an inability of the embryo to respond to the hormone. However, limited access to water may be more crucial for preventing germination on the mother plant in ABA-deficient mutants of *Arabidopsis thaliana* (Karssen et al., 1983).

Abscisic acid-deficient mutants of tomato undergo viviparous germination when ripe fruits remain attached to the mother plant (Groot, 1986). However, this viviparous condition is not due to a lower endogenous ABA level but rather to a relative insensitivity (as compared to wild-type seeds) to the low osmotic potential of the surrounding fruit tissues. Perhaps the different levels of ABA that are present at the various stages during seed development induce a different sensitivity to osmotic stress, which permits vivipary under appropriately hydrated conditions.

The germination of many mature, nondormant embryos can be inhibited by exogenous application of ABA (Milborrow, 1974). Exogenous ABA also appears to specifically inhibit the rise in activities of several enzymes (e.g., α-amylase, isocitrate lyase, and endo-β-mannanase) involved in post germinative breakdown of reserves (Dommes & Northcote, 1985; Halmer & Bewley, 1979; Ho & Varner, 1976). Abscisic acid influences transcription in cereal aleurone layers. Jacobsen and Beach (1985), using isolated nuclei from the aleurone layer of mature barley, demonstrated that ABA not only prevents GA-promoted accumulation of α-amylase and other gene transcripts, but also prevents the GA-suppression of total and ribosomal transcripts. It is not known if ABA plays any part in the control of α-amylase formation and secretion in the normal postgerminative mobilization processes of an intact (mature) seed. Under normal conditions, one would expect that it does not, since the level of ABA in mature barley seeds (and other cereals such as wheat) is low. However, the possibility exists that concentrations in the aleurone layer cells lie in the physiologically active range (Black, 1983). Also, ABA increases in the endosperm in response to water stress in germinated barley (J.V. Jacobsen, 1986, personal communication) and, hence, it may play some inhibitory role under adverse (stress) conditions.

King (1976) has postulated that accumulation of ABA during development of wheat seeds prevents precocious germination of the embryo and the premature hydrolysis of starch reserves of the morphologically complete but still immature seeds. The production of α-amylase by mature cereal embryos accompanies their germination but in immature isolated embryos, a close association between germination and α-amylase production is not apparent (Cornford et al., 1987). Thus, the suppression of enzyme formation in the intact immature seed is not simply due to the absence of germinative growth. Dure (1975) and Ihle and Dure (1972) originally suggested that the role of ABA in preventing vivipary of the immature cotton embryo is to suppress the translation of certain mRNAs—those for the proteins synthesized just after germination. But more recent experiments have offered alternative explanations (Radin & Trelease, 1976). Investigations on the effects of exogenous ABA on mature seeds have led to some understanding of the mode of action of the inhibitor on seed germination. Such studies indicate that ABA does not extend the lag period preceding radicle elongation but instead acts upon extension growth itself (i.e., ABA inhibits at the point of incipient radicle elongation) (Bex, 1972; Karssen, 1976; Schopfer et al., 1979). A mechanism involving restricted water uptake (or an interference with events connected with water uptake) has been invoked to explain the potency of the inhibitor on radicle emergence itself (Schopfer & Plachy, 1984, 1985).

In nondormant seeds, levels of endogenous ABA generally decline sharply after the seed reaches maximum dry or fresh weight, around the onset of maturation drying; levels at maturity are extremely low (Table 3-3). This decline in ABA during the later maturation stages may release the seed from the constraints of development and allow germination to occur (see subsequent discussion). In dormant seeds, the germination process is blocked; radicle emergence is prevented in a seed that is fully imbibed and metabolizing. Dormancy is usually overcome when the seed receives a specific stimulus such as light or low temperature (cold stratification); in some seeds, it is lost with the passage of time in the dry state (after ripening). Following the termination of dormancy, seed germination is completed, usually under conditions different from those which (originally) triggered the release from dormancy. An inhibitor such as ABA is an obvious candidate for an agent that imposes the block to germination (reviewed in Black, 1983; Walton, 1980/81). However, to date, there is no consistent evidence that seeds are forced into dormancy and maintained in this state by the action of ABA (Black, 1983). Correlations between the content of ABA and the onset of dormancy are generally poor, both during development in one particular species, i.e., apple (*Pyrus malus* L.) or when comparisons are made between seeds that do and do not become dormant (e.g., *Avena* spp.) (Balboa-Zavala & Dennis, 1977; Berrie et al., 1979). Similarly, no acceptable generalization can be made that dormant seeds are characterized by a high ABA status or that dormancy is broken when the ABA level falls (Black, 1983). However, suggestive evidence that changes in ABA sensitivity may be involved comes from studies of preharvest sprouting in cereals. Preharvest sprouting (unlike vivipary) generally occurs in mature, dried seeds still held on the ear of the mother plant when they become wetted by rain or high humidities (Black, 1983). Susceptibility to preharvest sprouting and the duration of seed dormancy varies between wheat cultivars (Reddy et al., 1985). Maturing seeds removed from a sprouting-susceptible cultivar and placed in water will easily germinate; in contrast, those removed from a sprouting-resistant cultivar do not. Differences in preharvest sprouting susceptibility also occur in the seeds of these cultivars of wheat when grown under field conditions (Walker-Simmons, 1987). Such differences in cultivar susceptibility have been used to examine the role of ABA levels and sensitivity responses in wheat embryo germination (Koch et al., 1982; Walker-Simmons, 1987). The amount of ABA in the embryos of the sprouting-susceptible cv. Greer is only about 25% lower than in those of the sprouting-resistant cv. Brevor. However, there are much larger differences between these cultivars in sensitivity to ABA, as measured by the capability of ABA to block embryo germination. This difference was particularly evident upon maturation drying; whereas embryos of the sprouting-susceptible cultivar lost sensitivity to ABA as the seed entered the desiccation stage, those from the sprouting-resistant cultivar did not (Walker-Simmons, 1987).

LIMITATIONS TO OUR UNDERSTANDING OF ABA AND ψ_s EFFECTS

Despite the evidence presented, we cannot conclude that the regulation of maturation and control of germinability of the developing embryo universally rests with ABA. Although there is considerable evidence that exogenous ABA inhibits germination processes while promoting developmental processes, evidence that endogenous ABA is involved in a similar manner is much less substantial. For example, there are several cases where no clear correlations can be found between ABA levels and changes in the ability of young seeds or embryos to germinate (Bewley & Black, 1985). Results in the literature from studies on the role of ABA in embryo development and germination are often confusing and contradictory, in part due to differences in the experimental conditions used to examine the effects of ABA in vitro (Davies & Bedford, 1982), as well as to differences between species in their response (or sensitivity) to ABA, and/or the importance of ABA with respect to a regulatory role. It is apparent that we need to know more about ABA in seeds, including its rate of synthesis or import into cells and tissues, its localization in cells as well as within tissues, a notion of what fraction of the total cellular hormone is active, and a better understanding of its mode of action at the molecular level. Then, perhaps we can hope to answer questions regarding its regulatory role in seeds in situ (Walton, 1980/81; Black, 1983). We also need to understand more about how sensitivity to ABA might change during embryo development—particularly during late maturation and desiccation (see subsequent discussion).

Less information is available on the regulatory role of restricted water uptake provided by a high osmolarity of surrounding seed tissues. It would be of particular interest to determine how ABA levels and sensitivity to a high ψ_s might interact.

The limitations of traditional approaches to study of regulatory factors in seed development (e.g., studies on hormonal content or in vitro responses to exogenous hormone) have already been emphasized. Recently, more decisive approaches have been used, including the study of mutants that are deficient in, or insensitive to, potential regulatory cues.

Finally, there is an inherent difficulty in studying a mechanism or process that is likely to involve a complex temporal sequence of events or cues that may be generated both internally (as part of an autonomous sequence) and externally (from surrounding tissues and other parts of the plant).

CONCLUSION

Desiccation is the normal terminal event in development and may itself promote, or lead to, a loss of control by tissues surrounding the embryo. Such a loss could be due to a reduction in the content and/or flow of ABA from the mother plant or surrounding seed tissues. Some evidence suggests that desiccation alters the hormonal balance of the seed: ABA breakdown may be enhanced by the natural drying process in cereals (King, 1976, 1979). Soybean can be induced to germinate precociously by slow drying that depletes the pool of endogenous ABA; premature drying of immature cereal grain leads to inducibility of α-amylase in response to added GA (Ackerson, 1984b; Evans et al., 1975). Preliminary results from our castor bean seed work also support the contention that premature drying causes a decline in endogenous ABA.

Alternatively, drying could decrease the sensitivity of the embryo to ABA, resulting in a loss of competence to respond to this hormone. For example, in wheat, rape, and soybean embryos, the decline in ABA levels in late development is accompanied by a corresponding decrease in tissue sensitivity to exogenous ABA (as measured by its capacity to inhibit germination and to maintain or enhance expression of developmental storage protein genes) (Eisenberg & Mascarenhas, 1985; Finkelstein et al., 1985; Quatrano, 1986; Williamson et al., 1985). It is not known whether this change in response to exogenous ABA is directly due to drying (e.g., via changes in receptor levels or conformation) or to a lowered endogenous ABA content. A change in sensitivity following (natural) drying is also suggested by several studies showing distinct differences between the metabolic responses of developing and mature embryos to ABA. Such is the case even when both are inhibited from germinating by ABA. For example, in cotton, synthesis of malate synthase (an enzyme associated with both late embryogeny and with post-germinative seedling growth) can either be stimulated or inhibited by ABA depending on the physiological maturity (and water status) of the isolated embryo (Choinski et al., 1981). A difference in response to ABA between immature and mature wheat embryos in relation to α-amylase (α-AMY$_2$ isozyme) production has also been noted (Cornford et al., 1987). Storage protein gene expression remains responsive to ABA in the precociously germinating rape embryo, unlike in mature tissue, which lacks this response to ABA (Crouch et al., 1985). Thus, ABA alone may not be sufficient to regulate activity of the storage protein genes; desiccation may be a prerequisite for the cessation of their expression. However, under some conditions, embryos isolated from mature seeds (which have undergone the natural drying process) may still be capable of responding to ABA in a "developmental" manner—i.e., by enhancing storage protein syntheses. For example, high levels of exogenous ABA can maintain E_m mRNA stability and increase the level of globulin mRNA (via its synthesis) in cultured mature embryos of wheat (Quatrano, 1986; Quatrano et al., 1986; Williamson et al., 1985). It has been claimed that ABA causes a re-induction of storage protein gene expression in mature embryos of mustard (*Sinapis alba* L.) (Fischer et al.,

1987). However, the possibility that ABA simply maintains or increases the stability of conserved storage protein messages (i.e., those that escape desiccation and are present in the mature dry seed) was not evaluated. But the "switch" from developmental-to germination- and growth-associated processes brought about by premature (or natural) drying of whole seeds may not be mediated solely through an altered sensitivity or response of the embryo to ABA; a decline in endogenous ABA may also be required. For example, embryos isolated from prematurely dried castor bean seeds exhibit a response to ABA that is intermediate between the developmental response and that characteristic of an embryo from a mature seed that has undergone the natural drying process (Kermode & Bewley, 1987b).

Finally, it is noteworthy that the use of isolated embryos for studying factors involved in the termination of development has certain limitations. The natural drying process occurs when the embryo is in an intact (i.e., whole seed) state and a series of interactions between the embryo and the surrounding seed tissues may occur—not only during drying itself, but also during the subsequent imbibition event—when the actual switch in metabolism takes place. Hence, the embryo may never acquire a complete independence from the seed environment even following drying. Both desiccation and rehydration of the embryo within the whole seed (i.e., in the presence of the surrounding seed tissues) may be a prerequisite for the complete cessation of developmental metabolism and switch towards a germination and growth regime.

REFERENCES

Ackerson, R.C. 1984a. Regulation of soybean embryogenesis by abscisic acid. J. Exp. Bot. 35:403–413.

——. 1984b. Abscisic acid and precocious germination in soybeans. J. Exp. Bot. 35:414–421.

Adams, C.A., M.C. Fjerstad, and R.W. Rinne. 1983. Characteristics of soybean seed maturation: Necessity for slow dehydration. Crop Sci. 23:265–267.

Ammirato, P.V. 1977. Hormonal control of somatic embryo development from cultured cells of caraway. Interactions of abscisic acid, zeatin, and gibberellic acid. Plant Physiol. 59:579–586.

Armstrong, C., M. Black, J.M. Chapman, H.A. Norman, and K. Angold. 1982. The induction of sensitivity to gibberellin in aleurone tissue of developing wheat grains. I. The effect of dehydration. Planta 154:573–577.

Balboa-Zavala, O., and F.G. Dennis. 1977. Abscisic acid and apple seed dormancy. J. Am. Soc. Hortic. Sci. 102:633–637.

Barratt, D.H.P. 1986. Modulation by abscisic acid of storage protein accumulation in *Vicia faba* L. cotyledons cultured in vitro. Plant Sci. 46:159–167.

Berrie, A.M.M., D. Buller, R. Don, and W. Parker. 1979. Possible role of volatile fatty acids and abscisic acid in the dormancy of oats. Plant Physiol. 63:758–764.

Bewley, J.D., and M. Black. 1985. Seeds. Physiology of development and germination. Plenum Publ. Corp., New York.

Bex, J.H.M. 1972. Effects of abscisic acid on oxygen uptake and RNA synthesis in germinating lettuce seeds. Acta Bot. Neerl. 21:203–210.

Black, M. 1983. Abscisic acid in seed germination and dormancy. p. 334–363. *In* F.T. Addicott (ed.) Abscisic acid. Praeger Publ., New York.

Braun, J.W., and A.A. Khan. 1975. Endogenous ABA levels in germinating and non-germinating lettuce seeds. Plant Physiol. 56:731–733.

Bray, E.A., and R.N. Beachy. 1985. Regulation by ABA of β-conglycinin expression in cultured developing soybean cotyledons. Plant Physiol. 79:746–750.

Choinski, J.S., and R.N. Trelease. 1978. Control of enzyme activities in cotton cotyledons during maturation and germination. Plant Physiol. 62:142-145.

----, ----, and D.C. Doman. 1981. Control of enzyme activities in cotton cotyledons during maturation and germination. III. In vitro embryo development in the presence of abscisic acid. Planta 152:428-435.

Cornford, C.A., M. Black, J.M. Chapman, and D.C. Baulcombe. 1986. Expression of α-amylase and other gibberellin-regulated genes in aleurone tissue of developing wheat grains. Planta 169:420-428.

----, ----, J. Daussant, and K.M. Murdoch. 1987. α-Amylase production by premature wheat (*Triticum aestivum* L.). J. Exp. Bot. 38:277-285.

Creelman, R.A., and J.A.D. Zeevaart. 1981. The effect of carotenoid inhibitors on abscisic acid production in corn seedlings. p. 123. *In* Annual report MSU-DOE Plant Res. Lab., East Lansing, MI.

Crouch, M.L. 1982. Non-zygotic embryos of *Brassica napus* L. contain embryo-specific storage proteins. Planta 152:520-524.

----, and I.M. Sussex. 1981. Development and storage-protein synthesis in *Brasicca napus* L. embryos in vivo and in vitro. Planta 153:64-74.

----, K. Tenbarge, A. Simon, R. Finkelstein, S. Scofield, and L. Solberg. 1985. Storage protein mRNA levels can be regulated by abscisic acid in *Brassica* embryos. p. 555-566. *In* L. van Vloten-Doting et al. (ed.) Molecular form and function of the plant genome. NATO Advanced Studies Institute Proceedings. Plenum Publ. Corp., New York.

Dasgupta, J. and J.D. Bewley. 1982. Desiccation of axes of *Phaseolus vulgaris* during development causes a switch from a developmental pattern of protein synthesis to a germination pattern. Plant Physiol. 70:1224-1227.

----, ----, and E.C. Yeung. 1982. Desiccation-tolerant and desiccation-intolerant stages during development and germination of *Phaseolus vulgaris* seeds. J. Exp. Bot. 33:1045-1057.

Davies, D.R., and I.D. Bedford. 1982. Abscisic acid and storage protein accumulation in *Pisum sativum* embryos grown in vitro. Plant Sci. Lett. 27:377-343.

Dommes, J., and D.H. Northcote. 1985. The action of exogenous abscisic acid and gibberellic acids on gene expression in germinating castor beans. Planta 165:513-521.

Dooner, H.K. 1985. *Viviparous-1* mutation in maize conditions pleiotropic enzyme deficiencies in the aleurone. Plant Physiol. 77:486-488.

Duchartre, M.P. 1852. Note sur la germination des céréales récoltées avant leur maturité. C. R. Hebd. Seances Acad. Sci. 35:940-942.

Dure, L.S. III. 1975. Seed formation. Annu. Rev. Plant Physiol. 26:259-278.

----, and G.A. Galau. 1981. Developmental biochemistry of cottonseed embryogenesis and germination. XIII. Regulation of biosynthesis of principle storage proteins. Plant Physiol. 68:187-194.

----, ----, and S. Greenway. 1980. Changing protein patterns during cotton cotyledon embryogenesis and germination as shown by in vivo and in vitro synthesis. Isr. J. Bot. 29:293-306.

----, S. Greenway, and G.A. Galau. 1981. Developmental biochemistry of cottonseed embryogenesis and germination: Changing messenger ribonucleic acid population as shown by in vitro and in vivo protein synthesis. Biochemistry 20:4162-4168.

Eeuwens, C.J., and W.W. Schwabe. 1975. Seed and pod wall development in *Pisum sativum* L. in relation to extracted and applied hormones. J. Exp. Bot. 26:1-14.

Eisenberg, A.J., and J.P. Mascarenhas. 1985. Abscisic acid and the regulation of the synthesis of specific seed proteins and their messenger RNAs during culture of soybean embryos. Planta 166:505-514.

Ellis, R.H., T.D. Hong, and E.H. Roberts. 1987. The development of desiccation-tolerance and maximum seed quality during seed maturation in six grain legumes. Ann. Bot. 59:23-29.

Evans, M., M. Black, and J.M. Chapman. 1975. Induction of hormone sensitivity by dehydration is the one positive role for drying in cereal seed. Nature (London) 258:144-145.

Finkelstein, R.R., and M.L. Crouch. 1984. Precociously germinating rapeseed embryos retain characteristics of embryogeny. Planta 162:125-131.

----, and M.L. Crouch. 1986. Rapeseed embryo development in culture on high osmoticum is similar to that in seeds. Plant Physiol. 81:907-912.

----, K.M. Tenbarge, J.E. Shumway, and M.L. Crouch. 1985. Role of ABA in maturation of rapeseed embryos. Plant Physiol. 78:630-636.

Fischer, W., R. Bergfeld, and P. Schopfer. 1987. Induction of storage protein synthesis in embryos of mature plant seeds. Naturwissenschaften 74:86-88.

Fong, F., J.D. Smith, and D.E. Koehler. 1983. Early events in maize seed development. 1-Methyl 1-3-phenyl-5-(3-[trifluoromethyl] phenyl)-4-(1-H)-pyridinone induction of vivipary. Plant Physiol. 73:899–901.

Galau, G.A., and L. Dure III. 1981. Developmental biochemistry of cottonseed embryogenesis and germination: Changing mRNA population as shown by reciprocal heterologous cDNA-mRNA hybridization. Biochemistry 20:4169–4178.

----, D.W. Hughes, and L. Dure III. 1986. Abscisic acid induction of cloned cotton late embryogenesis-abundant (Lea) mRNAs. Plant Mol. Biol. 7:155–170.

Goldberg, R.B., G. Hoschek, S.H. Tam, G.S. Ditta, and R.W. Breidenbach. 1981a. Abundance, diversity and regulation of mRNA sequence sets in soybean embryogenesis. Dev. Biol. 83:201–217.

Greenwood, J.S., D.J. Gifford, and J.D. Bewley. 1984. Seed development in *Ricinus communis* cv. Hale (castor bean). II. Accumulation of phytic acid in the developing endosperm and embryo in relation to the deposition of lipid, protein and phosphorus. Can. J. Bot. 62:255–261.

Groot, S.P.C. 1986. Hormonal regulation of seed development and germination in tomato. Studies on abscisic acid- and gibberellin- deficient mutants. Ph.D. diss., Univ. of Wageningen, Netherlands.

Halmer, P., and J.D. Bewley. 1979. Mannase production by the lettuce endosperm. Control by the embryo. Planta 144:333–340.

Harlan, H.V., and M.N. Pope. 1922. The germination of barley seeds harvested at different stages of growth. J. Hered. 13:72–75.

Hein, M.B., M.L. Brenner, and W.A. Brun. 1984. Concentrations of abscisic acid and indole-3-acetic acid in soybean seeds during development. Plant Physiol. 76:951–954.

Ho, D.T.H., and J.E. Varner. 1976. Response of barley aleurone layers to abscisic acid. Plant Physiol. 57:175–178.

Hsu, F.C. 1979. Abscisic acid accumulation in developing seeds of *Phaseolus vulgaris* L. Plant Physiol. 63:552–556.

Ihle, J.N., and L.S. Dure III. 1972. The developmental biochemistry of cottonseed embryogenesis and germination. III. Regulation of the biosynthesis of enzymes utilized in germination. J. Biol. Chem. 247:5048–5055.

Jacobsen, J.V., and L.R. Beach. 1985. Control of transcription of α-amylase and rRNA genes in barley aleurone protoplasts by gibberellin and abscisic acid. Nature (London) 316:275–277.

Karssen, C.M. 1976. Uptake and effect of abscisic acid during induction and progress of radicle growth in seeds of *Chenopodium album*. Physiol. Plant. 36:259–263.

----, D.L.C. Brinkhorst-vanderSwan, A.E. Breekland, and M. Koornneef. 1983. Induction of dormancy during seed development by endogenous abscisic acid: Studies on abscisic acid deficient genotypes of *Arabidopsis thaliana* (L.) Heynh. Planta 157:158–165.

Kermode, A.R., and J.D. Bewley. 1985a. The role of maturation drying in the transition from seed development to germination. I. Acquisition of desiccation-tolerance and germinability during development of *Ricinus communis* L. seeds. J. Exp. Bot. 36:1906–1915.

----, and ----. 1985b. The role of maturation drying in the transition from seed development to germination. II. Post-germinative enzyme production and soluble protein synthetic changes within the endosperm of *Ricinus communis* L. seeds. J. Exp. Bot. 36:1916–1927.

----, and ----. 1986a. Alteration of genetically regulated syntheses in developing seeds by desiccation. p. 59–84. *In* A.C. Leopold (ed.) Membranes, metabolism and dry organisms. Cornell Univ. Press, Ithaca, NY.

----, and ----. 1986b. The role of maturation drying in the transition from seed development to germination. IV. Protein synthesis and enzyme activity changes within the cotyledons of *Ricinus communis* L. seeds. J. Exp. Bot. 37:1887–1898.

----, D.J. Gifford, and J.D. Bewley. 1985. The role of maturation drying in the transition from seed development to germination. III. Insoluble protein synthetic pattern changes within the endosperm of *Ricinus communis* L. seeds. J. Exp. Bot. 36:1928–1936.

----, J. Dasgupta, S. Misra, and J.D. Bewley. 1986. The transition from seed development to germination: A key role for desiccation? Hortic. Sci. (spec. insert) 21:1113–1118.

----, and J.D. Bewley. 1987a. The role of maturation drying in the transition from seed development to germination. V. Responses of the immature castor bean embryo to isolation from the whole seed: A comparison with premature desiccation. J. Exp. Bot. 39:487–497.

----, and ----. 1987b. Regulatory processes involved in the switch from seed development to germination: Possible roles for desiccation and ABA. p. 59–76. *In* L. Monti and E. Porceddu (ed.) Drought resistance in plants. Physiological and genetic aspects. Commission of the European Communities, Luxemburg.

King, R.W. 1976. Abscisic acid in developing wheat grains and its relationship to grain growth and maturation. Planta 132:43-51.

----. 1979. Abscisic acid synthesis and metabolism in wheat ears. Aust. J. Plant Physiol. 6:99-108.

----. 1982. Abscisic acid and seed development. p. 157-181. In A.A. Khan (ed.) The physiology and biochemistry of seed development, dormancy and germination. Elsevier Biomedical Press, Amsterdam.

Koch, K.L., I.A. Tamas, and M.E. Sorrells. 1982. The role of abscisic acid and gibberellic acid in the control of preharvest sprouting of wheat. Hortic. Sci. 17:50.

Koornneef, M., G. Reuling, and C.M. Karssen. 1984. The isolation and characterization of abscisic acid-insensitive mutants of *Arabidopsis thaliana*. Physiol. Plant. 61:377-383.

Long, S.R., R.M.K. Dale, and I.M. Sussex. 1981. Maturation and germination of *Phaseolus vulgaris* embryonic axes in culture. Planta 153:405-415.

McDaniel, S., J.C. Smith, and H.J. Price. 1977. Response of viviparous mutants to abscisic acid. Maize Genet. News Lett. 51:85-86.

Milborrow, B.V. 1974. The chemistry and physiology of abscisic acid. Annu. Rev. Plant Physiol. 25:259-307.

Misra, S., and J.D. Bewley. 1985. Reprogramming of protein synthesis from a developmental to a germinative mode induced by desiccation of the axes of *Phaseolus vulgaris*. Plant Physiol. 78:876-882.

Neill, S.J., R. Horgan, and A.D. Parry. 1986. The carotenoid and abscisic acid content of viviparous kernels and seedlings of *Zea mays* L. Planta 169:87-96.

----, ----, and A.F. Rees. 1987. Seed development and vivipary in *Zea mays* L. Planta 171:358-364.

Norman, H.A., M. Black, and J.M. Chapman. 1982. Induction of sensitivity to gibberellic acid in aleurone tissue of developing wheat grains. II. Evidence for temperature-dependent membrane transitions. Planta 154:578-586.

Obendorf, R.L., and S.H. Wettlaufer. 1984. Precocious germination during in vitro growth of soybean seeds. Plant Physiol. 76:1023-1028.

Prevost, I., and M.Th. LePage-Degivry. 1985. Inverse correlation between ABA content and germinability throughout the maturation and the in vitro culture of the embryo of *Phaseolus vulgaris*. J. Exp. Bot. 36:1457-1464.

Quatrano, R.S. 1986. Regulation of gene expression by abscisic acid during angiosperm embryo development. Oxford Surv. Plant Mol. Cell Biol. 3:467-477.

----, J. Litts, G. Colwell, R. Chakerian, and R. Hopkins. 1986. Regulation of gene expression in wheat embryos by ABA: Characterization of cDNA clones for the E_m and putative globulin proteins and localization of the lectin wheat germ agglutinin. p. 127-136. In L.M. Shannon and M.J. Chrispeels (ed.) Molecular biology of seed storage proteins and lectins. Am. Soc. Plant Physiol., Baltimore.

Quebedeaux, B., P.B. Sweetser, and J.C. Rowell. 1976. Abscisic acid levels in soybean reproductive structures during development. Plant Physiol. 58:363-366.

Radin, J.W., and R.N. Trelease. 1976. Control of enzyme activities in cotton cotyledons during maturation and germination. Plant Physiol. 57:902-905.

Radley, M. 1979. The role of gibberellin, abscisic acid, and auxin in the regulation of developing wheat grains. J. Exp. Bot. 30:381-389.

Raghavan, V. 1986. Embryogenesis in angiosperms. A developmental and experimental study. Cambridge Univ. Press, Ithaca, NY.

Raikhel, N.V., and R.S. Quatrano. 1986. Localization of wheat-germ agglutinin in developing wheat embryos and those cultured in abscisic acid. Planta 168:433-440.

Reddy, L.V., R.J. Metzger, and T.M. Ching. 1985. Effect of temperature on seed dormancy of wheat. Crop Sci. 25:455-458.

Robichaud, C.S., J. Wong, and I.M. Sussex. 1980. Control of in vitro growth of viviparous embryo mutants of maize by abscisic acid. Dev. Genet. 1:325-330.

Schopfer, P., D. Bajracharya, and C. Plachy. 1979. Control of seed germination by abscisic acid. I. Time course of action in *Sinapis alba* L. Plant Physiol. 64:822-827.

----, and C. Plachy. 1984. Control of seed germination by abscisic acid. II. Effect on embryo water uptake in *Brassica napus*. Plant Physiol. 76:155-160.

----, and ----. 1985. Control of seed germination by abscisic acid. III. Effect on embryo growth potential (minimum turgor pressure) and growth coefficient (cell wall extensibility) in *Brassica napus* L. Plant Physiol. 77:676-686.

Schroeder, H.E. 1984. Effects of applied growth regulators on pod growth and seed protein composition in *Pisum sativum* L. J. Exp. Bot. 35:813-821.

Smith, J.D., S. McDaniel, and S. Lively. 1978. Regulation of embryo growth by abscisic acid in vitro. Maize Genet. Coop. New Lett. 52:107-108.

Stinissen, H.M., W.J. Peumans, and E. deLanghe. 1984. Abscisic acid promotes lectin biosynthesis in developing and germinating rice embryos. Plant Cell Rep. 3:55-59.

Sussex, I.M. 1975. Growth and metabolism of the embryo and attached seedling of the viviparous mangrove, *Rhizophora mangle*. Am. J. Bot. 62:948-953.

----, and R.M.K. Dale. 1979. Hormonal control of storage protein synthesis in *Phaseolus vulgaris*. p. 129-141. *In* I. Rubenstein et al. (ed.) The plant seed. Development, preservation and germination. Academic Press, New York.

Triplett, B.A., and R.S. Quatrano. 1982. Timing, localization and control of wheat germ agglutinin synthesis in developing wheat embryos. Dev. Biol. 91:491-496.

Van Onckelen, V., R. Caubergs, S. Horemans, and J.A. Degreef. 1980. Metabolism of abscisic acid in developing seeds of *Phaseolus vulgaris* L. and its correlation to germination and α-amylase activity. J. Exp. Bot. 31:913-920.

Walker-Simmons, M. 1987. ABA levels and sensitivity in developing wheat embryos of sprouting resistant and susceptible cultivars. Plant Physiol. 84:61-66.

Walton, D.C. 1980/81. Does ABA play a role in seed germination? Isr. J. Bot. 29:168-180.

Wellington, P.S. 1956. Studies on the germination of cereals. I. The germination of wheat grains in the ear during the development, ripening, and after-ripening. Ann. Bot. 20:105-120.

Williamson, J.D., R.S. Quatrano, and A.C. Cuming. 1985. E_m polypeptide and its messenger RNA levels are modulated by abscisic acid during embryogenesis in wheat. Eur. J. Biochem. 152:501-507.

Yeung, E.C, and D.C.W. Brown. 1982. The osmotic environment of developing embryos of *Phaseolus vulgaris*. Z. Pflanzenphysiol. 106:149-156.

4 Moisture as a Regulator of Physiological Reaction in Seeds

A. Carl Leopold
Boyce Thompson Institute
Ithaca, New York

Christina W. Vertucci
USDA-ARS
National Seed Storage Laboratory
Fort Collins, Colorado

Physiological activity is reduced as orthodox tissues become dried and resumes as the tissue is rehydrated. Since water functions in the cell in a variety of ways, the loss of activity when the system becomes dry may be due to a number of factors. Water is an important substrate in many reactions. Its removal can lead to reduced activity due to low substrate concentrations. Probably more important is the role of water as the solvent for most biochemical reactions. Loss of the solvent will reduce the diffusion rate of solute substrates to an active site (Acker, 1969; Duckworth, 1962). Water also affects the intramolecular motions of proteins that are essential for catalytic activity (Jaenicke, 1981). Finally, water effects on phospholipid structure may play an important role in membrane permeability (Crowe & Crowe, 1986). The extent to which activity is limited by dry conditions is critical in maintaining energy reserves within the organism and also for preventing the accumulation of deleterious compounds (Priestley, 1986; Bewley, 1979).

The survival of a seed in dry storage depends upon its moisture content more than on any other factor (Justice and Bass, 1978). This dependence in dry seeds might be attributed intuitively to the notion that physiological reactions may quantitatively increase as water content increases. Deteriorative reactions frequently proceed in the seed more readily if the moisture content is higher, and consequently the moisture condition would constitute a threat to longevity of survival. The impressive sensitivity of seed longevity to moisture content is illustrated in Fig. 4-1 (Justice & Bass, 1978) where seeds of soybean [*Glycine max* (L.) Merr.] readily retain germinability at 0.10 g H_2O g^{-1} dry wt. (dw)[1], but deteriorate rapidly at 0.16 or 0.22 g H_2O g^{-1}

---

[1] To convert (g H_2O g^{-1} dw) to (%), multiply by 100. Thus, 0.10 g H_2O g^{-1} dw = 10%.

Copyright © 1989 Crop Science Society of America, 677 S. Segoe Rd., Madison, WI 53711, USA. *Seed Moisture*, CSSA Special Publication no. 14.

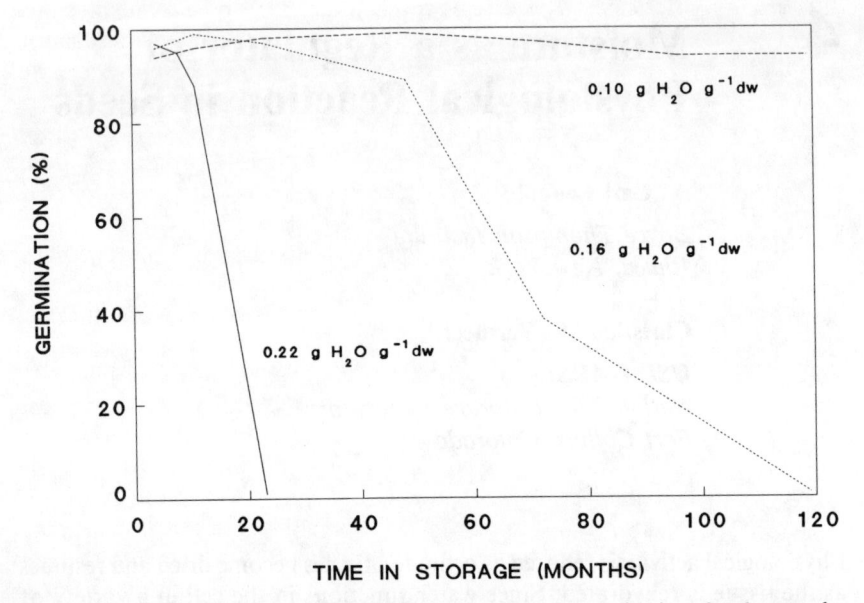

Fig. 4-1. The effect of moisture content and storage time on the germination of soybean seeds. Data show that at 10 °C, seeds stored are relatively stable at 0.10 g H$_2$O g^{-1} dw within a period of about 10 yr, while seeds stored at 0.22 g H$_2$O g^{-1} dw lose viability within 2 yr. Data from Justice and Bass (1978).

dw moisture content. From data such as these, one might infer that deteriorative reactions may proceed at the higher moisture levels, but that they may be restrained at the lower levels. Recent advances in the understanding of the thermodynamic status of water in seeds and its relationship to reactions, however, has led to a considerable departure from that intuitive notion. In fact, some reactions occur more readily in the dry state than in the wet, and the nature of the reactions that occur will vary with the water content.

The precise relationship between metabolic activity in seeds and moisture contents is, to a large degree, unknown. Many studies of the rapid resumption of physiological activity with imbibition are based on timed measurements and the actual moisture contents are unknown. For example, increased lipoxidase activity in soybean seed was noted within a minute of imbibition (Boveris et al., 1983) and mitochondrial activity in seeds developed within hours (Nawa & Asahi, 1973; Wilson & Bonner, 1971; Opik & Simon, 1963; Morohashi, 1978). These reports do not help to clarify the precise role of water content in controlling physiological activity in seeds. The intent of this chapter is to provide a brief review of the thermodynamic status of water in dry seeds, and to describe examples of reactions that apparently occur within the various regions defined by the thermodynamic properties of water.

A classic study of physiological activity in an anhydrobiotic organism was done by Clegg and co-workers using *Artemia* cysts. Clegg (1978) separated the "resting" state of cysts into three levels of metabolic activity that correspond to hydration: an ametabolic state where no enzyme-mediated reactions could occur, a state with intermediate metabolism in which simple reactions could occur and finally a fully active metabolic state where integrated processes like mitochondrial electron transport was observed. Clegg related the three levels of physiological activity with the physical condition of the water within the cysts. It is implied, therefore, that metabolic activity cannot occur at the lowest hydration levels but can occur in the absence of free water.

With the moisture contents at which seeds are ordinarily stored, water exists as bound water. The term *bound water* refers to water associated with a macromolecular surface and is sufficiently structured so that its thermodynamic properties differ from free or bulk water; most notably, it is not readily freezable. Illustrating the bound nature of water are data in Fig. 4-2, where soybean seeds had been equilibrated to various moisture contents and then exposed to $-65\,°C$, after which they were brought back to room temperature, hydrated, and germinated (Leopold & Vertucci, 1986). It is clear that the germinability was lost at moisture contents above about 0.35 g H_2O g^{-1} dw, but the seeds at lower moisture contents survived freezing in terms of retaining germinability and yielding little leakage of solutes into the germination medium. From such experiments and experiments using differential thermal analysis and differential scanning calorimetry (Vertucci & Leopold, 1986) that showed no freezing events in soybean at water contents

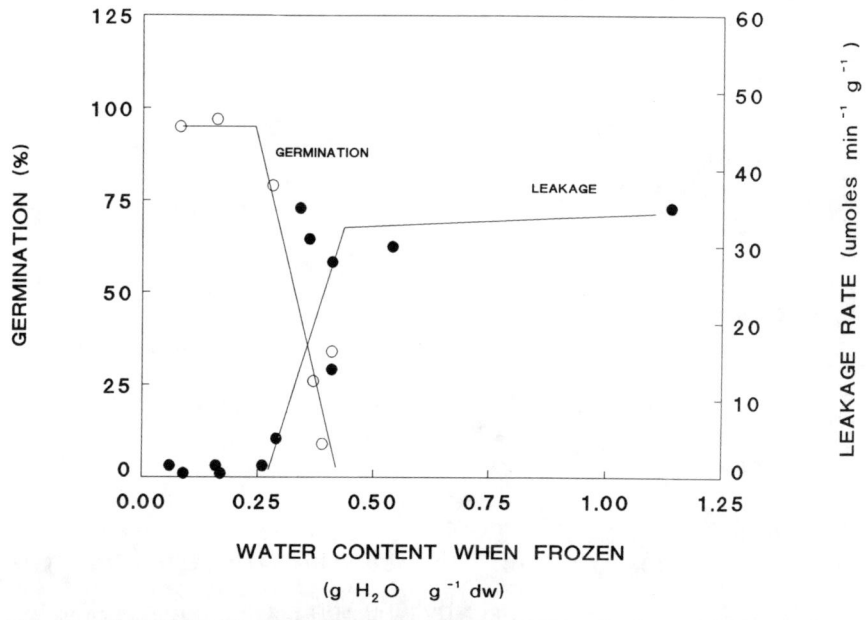

Fig. 4-2. The effects of freezing of soybean seeds to $-65\,°C$ on seed germination and electrolyte leakage. Exposure to low temperatures results in germinability loss and the increase of solute leakage when moisture contents were > 0.25 to 0.35 g H_2O g^{-1} dw. Data from Leopold and Vertucci (1986).

<0.35 g H$_2$O g^{-1} dw, it is concluded that the water in soybean at lower moisture contents is structured and hence it is not free water.

DEFINING SEED MOISTURE REGIONS

Studies of water mobility and thermodynamics in protein samples and in seeds have contributed to the notion that there are at least three types of bound water (Rupley et al., 1983; Rockland, 1969). The type of bound water is determined by the strength with which it is sorbed to a macromolecular surface. Type 1 water is believed to "chemi-sorb" to macromolecules through ionic bonding. Type 2 water condenses over the hydrophilic sites of macromolecules, and type 3 water forms bridges over hydrophobic sites (Rupley et al., 1983). Consequently, type 1 water is bound extremely tightly, type 2 is bound weakly and type 3 water is bound with negligible energy.

Within the moisture range in which water is bound, regions with different water-binding characteristics can be defined using moisture isotherms. These are measured by equilibrating seeds under various relative humidities (RH), and plotting the equilibrated moisture content against the RH (Vertucci & Leopold, 1984). Moisture isotherms for soybean seeds obtained at two temperatures are illustrated in Fig. 4-3. The curves indicate that there is a strong avidity for absorbing moisture in humidities below about 20% RH, a region of weaker avidity between 20 and 60% RH, and then a region of abundant water binding at humidities >60%.

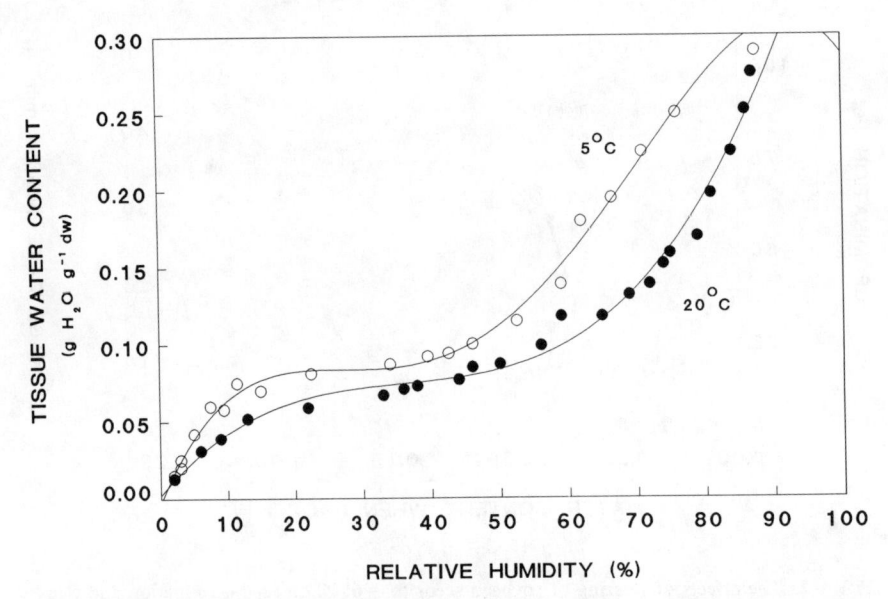

Fig. 4-3. Water sorption isotherms of ground soybean pellets at 5 and 20 °C, showing the reverse sigmoidal curve typically found for seeds. Relative humidity was controlled by saturated salt solutions. Data from Vertucci and Leopold (1984).

The affinity of water for seed components can be calculated from the isotherms in terms of the enthalpy of binding. Two different methods can be used. In the first, one can utilize the Clausius-Clapeyron equations[2] that compare the RH values at which similar wetting is observed at two temperatures. The calculated enthalpies of water binding for seeds of soybean, pea (*Pisum sativum* L.), and apple (*Malus domestica* L.) are plotted in Fig. 4-4 (Vertucci & Leopold, 1986). In each case, a region of large negative enthalpies of binding at low moistures, a region of higher enthalpies of binding at intermediate water contents, and finally a region of almost no energy of binding at higher moisture contents are seen. These three water-binding regions are similar to regions described previously for water absorption onto globular proteins (Rupley et al., 1983; Careri et al., 1979; D'Arcy & Watt, 1970; Bull, 1944) and *Artemia* cysts (Clegg, 1978).

The regions of water binding can also be calculated by the D'Arcy-Watt equation[3], which utilizes the characteristic three regions of moisture isotherms to define the thermodynamic properties of binding in each region (Vertucci & Leopold, 1987a). This equation is based on an algebraic separation of a nearly linear region in the middle of the moisture isotherms from the nonlinear regions above and below it. The equation allows one to calculate the values for water binding in each of the three binding regions. The binding affinities for the axes of five species of legume seeds are given in Table 4-1 from which one can see that the affinity for water in region 1, the region of lowest water content, is almost an order of magnitude greater than the affinity for water in region 2. In the five species, the axes have characteristically much higher water-binding properties than the cotyledons (Vertucci & Leopld, 1987a).

The experimental results shown so far should serve to emphasize that the water in dry seeds is structured and nonfreezable. Structured water can be separated into three regions with distinctive affinities, the region with the lowest water contents has the highest affinity for water. Thus, we can describe a model of hydration onto seed macromolecular surfaces starting with a dry surface and adding moisture incrementally. Water in region 1 has limited

---

[2]The Clausius-Clapeyron equation:

$$H = (R\ T_1\ T_2)/(T_2 - T_1)\ [\ln(a_{w_1}/a_{w_2})]$$

where H = the differential enthalpy of hydration,
R = the gas constant,
T_1 and T_2 = the lower and higher temperatures, and
a_{w_1} and a_{w_2} = the relative vapor pressure at T_1 and T_2.

[3]The D'Arcy-Watt equation:

$$W = \{(KK'\ a_w/[(1 + K)\ a_w]\} + [c\ a_w] + \{(kk'\ a_w)/[(1 - k)\ a_w]\}$$

where W = the grams of water sorbed per gram sorbent
a_w = the relative vapor pressure
K and K' = the affinity and number of sites in water-binding region 1, respectively,
c = the number × the affinity of sites in water-binding region 2, and
k and k' = the affinity and number of sites in water-binding region 3.

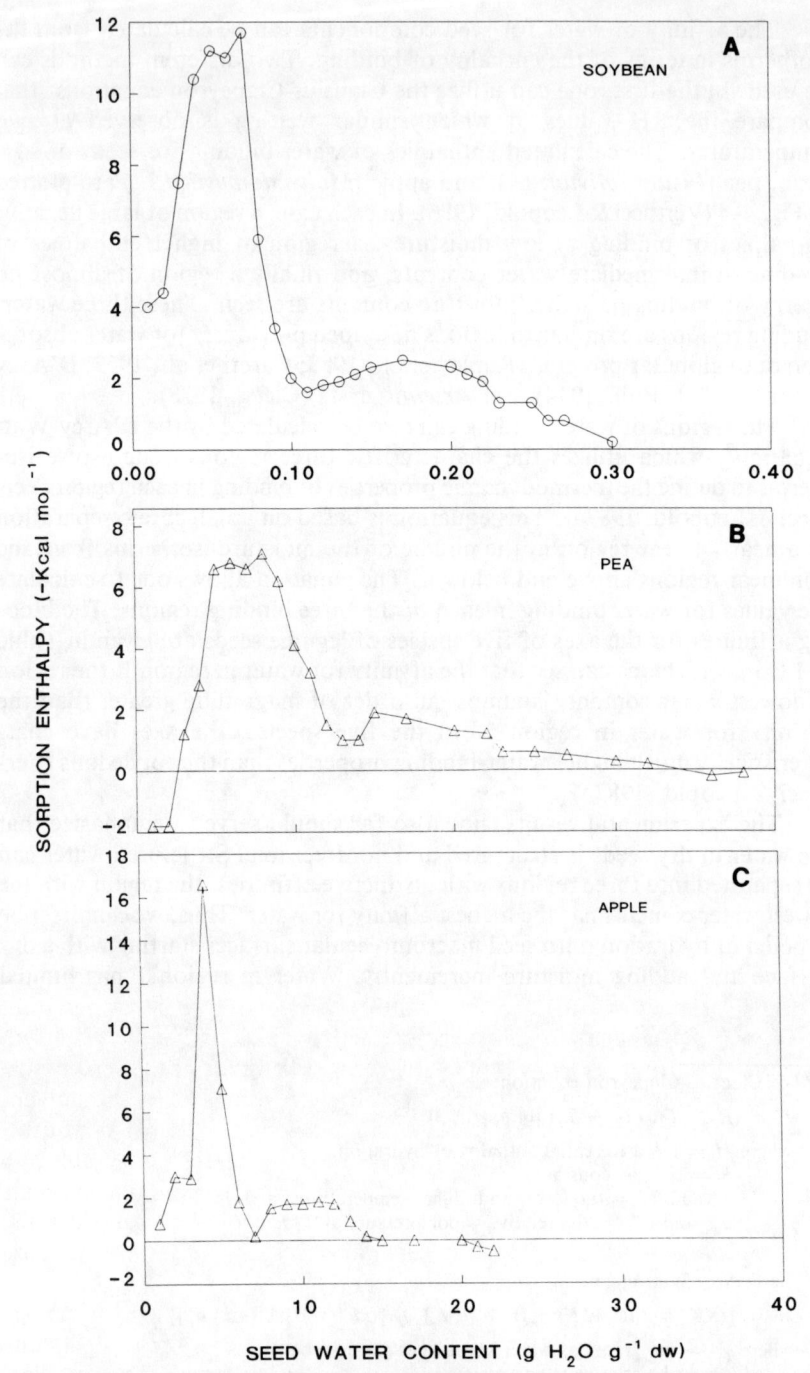

Fig. 4-4. Differential enthalpies of water sorption as a function of moisture content in soybean, pea, and apple embryos. Values are calculated from isotherms at 15 and 25 °C (soybean and pea) and 5 and 15 °C (apple) using the Clausius-Clapeyron equation. Data from Vertucci and Leopold (1986).

Table 4-1. The integral strength of water binding in three regions for axis tissue of five legume seed species at 5 °C. Calculated from moisture isotherms by D'Arcy-Watt equation (Vertucci & Leopold, 1987a).

Seed Species	Region 1	Region 2	Region 3
	——kJ mol^{-1}——		water activity[†] (dimensionless)
Soybean [*Glycine max* (L.) Merr.]	14.0	2.0	1.01
Pea (*Pisum sativum* L.)	12.1	2.0	1.02
Cowpea (*Vigna sinensis* Endl.)	13.1	2.0	1.01
Fava bean (*Vicia faba* L.)	13.1	2.0	1.02
Peanut (*Arachis hypogaea* L.)	13.6	2.0	1.03

[†] The values of water activity in region 3 approximate 1.0, which suggest that water binding in this region is about 0 kJ mol^{-1}.

mobility and is believed to behave as a ligand rather than a solvent. Water begins to have solvent properties in region 2 when diffusion gradually becomes evident (Duckworth, 1962) and the properties of water begin to resemble those of the bulk solution. In region 3, water properties are similar to bulk water, although a few thermodynamic parameters show variations. The macromolecular surface is said to be fully wetted when no further changes in the properties of water are observed with further hydration (about 0.35 g H_2O g^{-1} dw in soybean seeds). With this preparation, we can examine an array of physiological reactions in dry seeds for possible association with the three regions of water binding.

REACTIONS IN DRY SEEDS

Nonenzymatic Reactions

Experiments have been carried out to probe nonenzymatic oxidations in soybean seeds at moisture contents of 0.05 and 0.19 g H_2O g^{-1} dw (Priestley et al., 1985b). In pure O_2 atmospheres and at high temperatures (105 °C), which will destroy most enzyme activity, autooxidations occurred, principally causing discoloration of the seeds. The lipids in the seed tissue, however, were somehow protected from the oxidation. Even the polyunsaturated fatty acids were essentially unscathed by this harsh treatment. However, extracted oils treated similarly were particularly susceptable and were promptly oxidized. From these experiments, it seems that nonenzymatic oxidations can occur in the seed, but fatty acids appear to be markedly resistant to this type of attack. Evidence from model systems suggest that autooxidations are favored at higher moisture contents (Chan, 1987).

Another type of oxidative activity that might occur in dry seeds in storage is due to attack by free radicals. Measurements of the content of organic free radicals in soybean axes and cotyledons have been carried out using electron spin resonance (ESR), and the results indicate that the frequency of such species is markedly elevated at moisture contents below 0.12 to 0.15 g H_2O

g^{-1} dw (Fig. 4-5) (Priestley et al., 1985a). Damage to kenaf (*Hibiscus cannabinus* L.) seeds by free radicals is dependent on hydration; seeds within the first and third hydration region are damaged by free radical-producing radiation, while seeds in the second hydration region are somehow protected (Mahama & Silvy, 1982). High-temperature/high-humidity treatments of seeds induces an increasing tendency toward free-radical accumulation when seeds are subsequently dried to 0.08 g H_2O g^{-1} dw (Priestley et al., 1985a). These data suggest that free radicals may play a role in deteriorative processes in dry seeds, and that the ability to scavenge radicals is dependent on the moisture level and seed vigor.

Enzyme Reactions

Most of the studies of enzyme activity as a function of moisture content involved model systems of enzymes in which activity was not observed below a minimum moisture content (about 0.08 g H_2O g^{-1} dw), and further additions of water caused much greater activity (Acker, 1969; Careri et al., 1980; Rupley et al., 1983). The activity of lipolytic enzymes was detectable at low water activities and appeared to be controlled predominantly by the fluidity of the lipid milieu (Labuza, 1980; Brockmann & Acker, 1977; Acker, 1969). The activity of water-soluble enzymes increased monotonically as RH increased above 25% (approximately, the onset of the second region of hydration). In contrast, the activity of enzymes with lipolytic functions increased with RH in a reverse sigmoidal fashion parallel to the sorption isotherm (Acker, 1969). In whole tissue, lipolytic functions appeared to decrease at high-moisture levels (region 3) (Brockmann & Acker, 1977; Labuza, 1980). Clearly, bulk water is not necessary for activity of all enzymes (Potthast, 1978), but the presence of clustered water facilitates enzymic reactions. Because oxidases are affected in a similar manner as hydrolases (Acker, 1969), the presence of water as a substrate is probably not the most important limiting factor for enzyme activity. Acker (1969) contends that mobility of the substrate is most important. It is also possible that the presence of type 2 water allows the intramolecular motions necessary for the catalytic activity of proteins.

In soybean, when the moisture content was within region 2, there was a low, but measurable, uptake of O_2 that yielded an RQ of <0.5 (Vertucci & Leopold, 1986, 1987b). It is believed that oxidative reactions at these water contents do not involve metabolic electron transport (Vertucci & Leopold, 1986, 1987b; Lynch & Clegg, 1986), but rather are the consequences of enzymes or stochastic processes. Since autooxidations of unsaturated lipids appear to be limited in intact tissues (Priestley et al., 1985b or at ambient temperatures (Boveris et al., 1983), and since lipoxidase can function at low water activities (Boveris et al., 1983; Brockmann & Acker, 1977; Labuza, 1980), it was surmised that lipoxidase activity may be contributing to the measured O_2 consumption. Lipoxidase is controlled by phytochrome, red light being inhibitory (Oelze-Karow & Mohr, 1972; Schoper, 1977). Pursuing the idea that the O_2 uptake measured in the second water-binding region

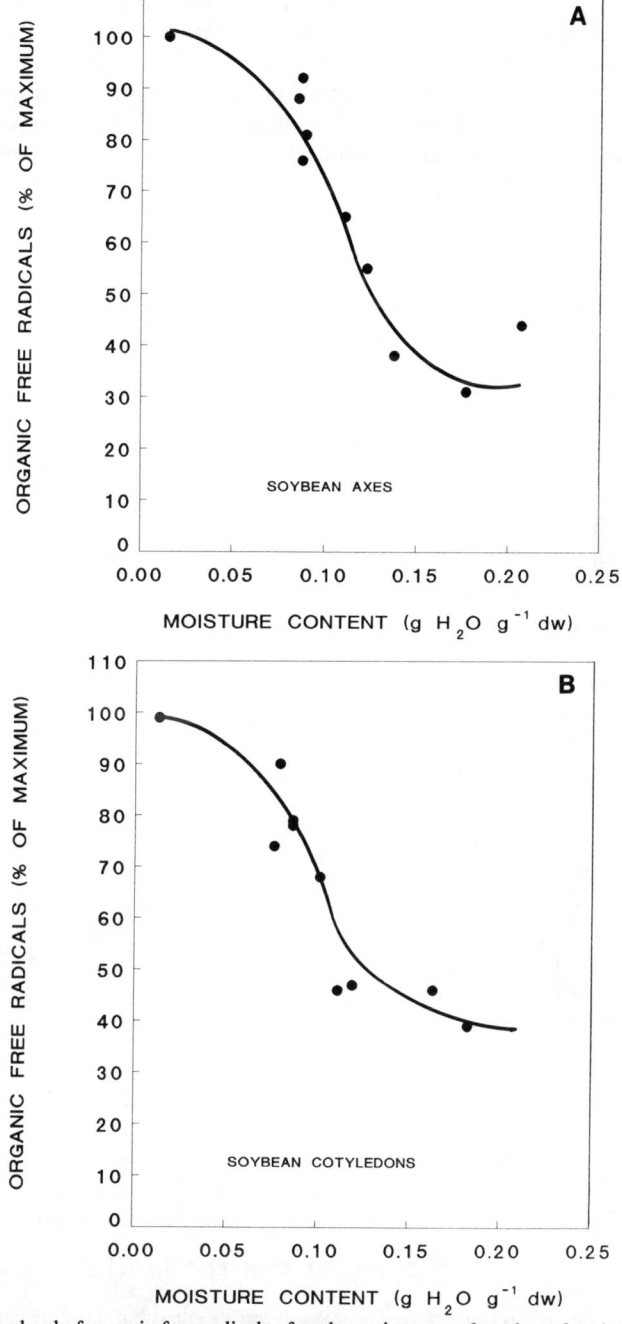

Fig. 4-5. The level of organic-free radicals of soybean tissue as a function of moisture content. Powdered seed material was equilibrated to different water contents in controlled relative humidity chambers. Free-radical content was subsequently determined by electron spin resonance. Data from Priestley et al. (1985a).

was due to lipoxidase activity, the effect of light on O_2 uptake rates was measured (Vertucci & Leopold, 1987b). The rate was inhibited by light at water contents between 0.08 and 0.25 g H_2O g^{-1} dw (Fig. 4-6). Based on such data and evidence suggesting that lipoxidase activity is limited at high-moisture levels (Labuza, 1980; Brockmann & Acker, 1977), we speculate that the oxidative processes in seeds with moistures in the second region of water binding may be enzymic.

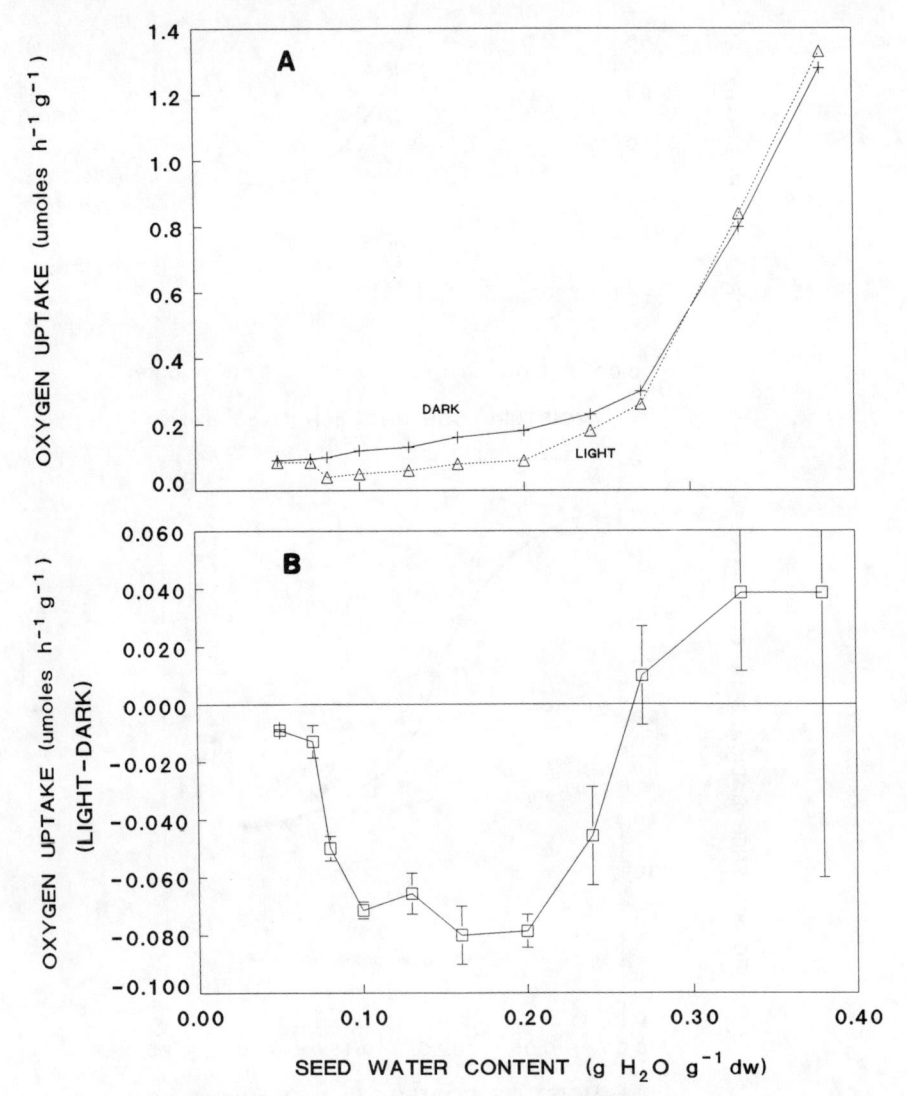

Fig. 4-6. The rate of O_2 consumption in soybean seeds at different moisture contents under dark and lighted conditions. Rates are measured at 25 °C. (A) Rates in dark or light as a function of water content. (B) The difference between the rates in light vs. dark as a function of water content. The error bar represents the standard error of four replicates. It is ±0.0125 for water contents <0.24 g H_2O g^{-1} dw. Data from Vertucci and Leopold (1987b).

Integrated Metabolism

Enzymes that only require type 2 water for activity have catabolic functions. Anabolic activity is first observed at much higher water contents, when type 3 water is present (Clegg, 1978). When radioactive tracers were applied, the labels were incorporated into proteins and other macromolecules of *Artemia, Pinus ponderosa* pollen, and lichens when organisms were incubated at moisture content >0.65 g H_2O g^{-1} dw (*Artemia*) or 0.25 g H_2O g^{-1} dw (plant materials) (Clegg, 1978; Wilson et al., 1979; Cowan et al., 1979).

The moisture content necessary for respiratory activity has received much attention. Examination of O_2 uptake rates as a function of water content was made for four species in a Gilson respirometer (Fig. 4-7) (Vertucci & Leopold, 1986). The minor amounts of CO_2 evolution or O_2 uptake that have been noted at moisture contents comparable to the second hydration level (Ching, 1961; Bartholomew & Loomis, 1967; Vertucci & Leopold, 1984, 1986, 1987b) are probably enzymic reactions (Priestley, 1986; Bartholomew & Loomis, 1967; Vertucci & Leopold, 1986, 1987b). There is a large increase in the rate of O_2 consumption when water content increases to levels >0.27, 0.24, 0.20, and 0.12 g H_2O g^{-1} dw for pea, soybean, maize (*Zea mays* L.), and apple, respectively (Fig. 4-7). It is believed that at these moisture contents, mitochondrial respiration is potentiated. Supporting this hypothesis is the observation that the respiratory quotient in soybean abruptly increases from <0.5 to 1.0 at 0.24 g H_2O g^{-1} dw (Vertucci & Leopold, 1986). Also, although ATP could be consumed in the second hydration level, it could not

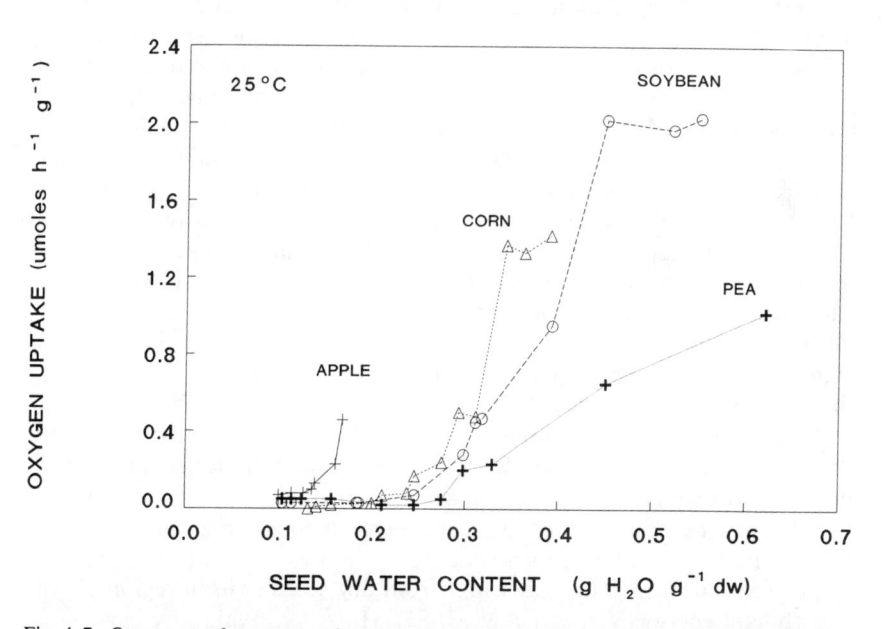

Fig. 4-7. Oxygen uptake rates as a function of water content from apple, maize, soybean, and pea. Measurements, expressed as mol O_2 h^{-1} g^{-1}, were taken in a Gilson respirometer at 25°C. Data from Vertucci and Leopold (1986).

be synthesized unless type 3 water was present (A.C. Leopold, 1984, unpublished data). Comparisons of the rate of O_2 consumption with moisture content (Fig. 4-7) and the enthalpy curves for water binding (Fig. 4-4) indicate that metabolic O_2 uptake becomes operable as type 3 water becomes available.

There is evidence that mitochondria in dry seeds are somewhat "incomplete," but that resumption of mitochondrial activity is rapid during imbibition (Nawa & Asahi, 1973; Wilson & Bonner, 1971; Opik & Simon, 1963; Morohashi, 1978). Initially, respiration is limited by inoperative electron transport and oxidative phosphorylation reactions (Nawa & Asahi, 1973; Wilson & Bonner, 1971; Opik & Simon, 1963; Morohashi, 1978). Increased O_2 uptake with hydration is associated with stronger interactions of mitochondrial proteins with the mitochondrial matrix (Nawa & Asahi, 1971; Wilson & Bonner, 1971; Opik & Simon, 1963; Morohashi, 1978).

The effect of water content on photosynthetic activity can be studied in pea seeds. Data from fluorescence-induction kinetics studies demonstrate that photosynthetic electron transport is functional within the second region of water binding (Vertucci et al., 1985). However, carbon fixation, measured by O_2 evolution in the light, was not possible until the tissue was fully wetted (>0.35 g H_2O g^{-1} dw) (Vertucci & Leopold, 1987b).

Light Absorption by Pigments

We have discussed examples of reactions that might occur in seeds either through enzymatic processes or autooxidations. These are thermal reactions, requiring inter- and intramolecular mobility. Energy requirements for oxidative activities are high as demonstrated by apparent activation energies that increase with decreasing moisture contents (Vertucci & Leopold, 1987b). A measure of the thermal energy of seed tissue and its relation to thermal-chemical reactions is found in comparisons of the free energy (ΔG), calculated from sorption isotherms (Vertucci & Leopold, 1987a) and the average kinetic energy of a molecule (kT, where k is the Boltzmann constant and T is temperature). When $\Delta G < -kT$, the amount of thermal energy in the system will be unfavorable for thermal reactions and as ΔG approaches $-kT$ spontaneous thermal reactions become increasingly possible. In pea and lettuce (*Lactuca sativa* L.), ΔG equals kT at 0.12 and 0.08 g H_2O g^{-1} dw, respectively (Vertucci & Leopold, 1987a; Vertucci et al., 1987). This amount suggests that thermal reactions are essentially excluded from seeds stored below these moisture contents. It would be interesting, then, to determine whether nonthermal reactions might take place.

The absorption of light energy by pigments, a nonthermal reaction, seems to be nominally affected by water content. Fluorescence excitation spectra of pea seeds demonstrate that Chl *a* from dry seeds (within region 1) will absorb light energy (Vertucci et al., 1985). However, qualitative changes in the excitation spectrum suggest that the transfer of the light energy to reaction centers is limited within region 1 (Vertucci et al., 1985).

Phytochrome, another plant pigment, can exist in dry seeds as the red absorbing form (Pr) or the far-red absorbing form (Pfr). A photoreaction is necessary to interconvert the pigments. Hsiao and Vidaver (1971) suggested that reversible conversion required at least 0.15 g H_2O g^{-1} dw. More recently, research has suggested that Pfr can be photoconverted at lower water contents (Kendrick & Russell, 1975; Bartley & Frankland, 1984). Experiments using lettuce seeds have indicated that indeed phytochrome conversion from the far-red to the red absorbing form can occur at any moisture content, even as low as 0.04 g H_2O g^{-1} dw (Vertucci et al., 1987) (Fig. 4-8). These results demonstrate that, whereas thermal reactions may be restricted in the lowest region of water binding, photoreactions are able to occur at any region of water binding.

This review of reactions that occur in seeds with different regions of hydration leads us to suggest some generalizations. First, enzyme activity is possible in the second hydration region and oxidations that occur in seeds at these water contents may be due either to enzymes or stochastic processes. Second, integrated metabolic activity is facilitated in seeds near the boundary between water-binding regions 2 and 3. We have emphasized that the seed with intermediate moisture contents may have effective protective devices serving to limit the extent of free radical-induced oxidations in dry storage. Reactions that do not require thermal energy, such as photochemical reactions can occur at all moisture levels.

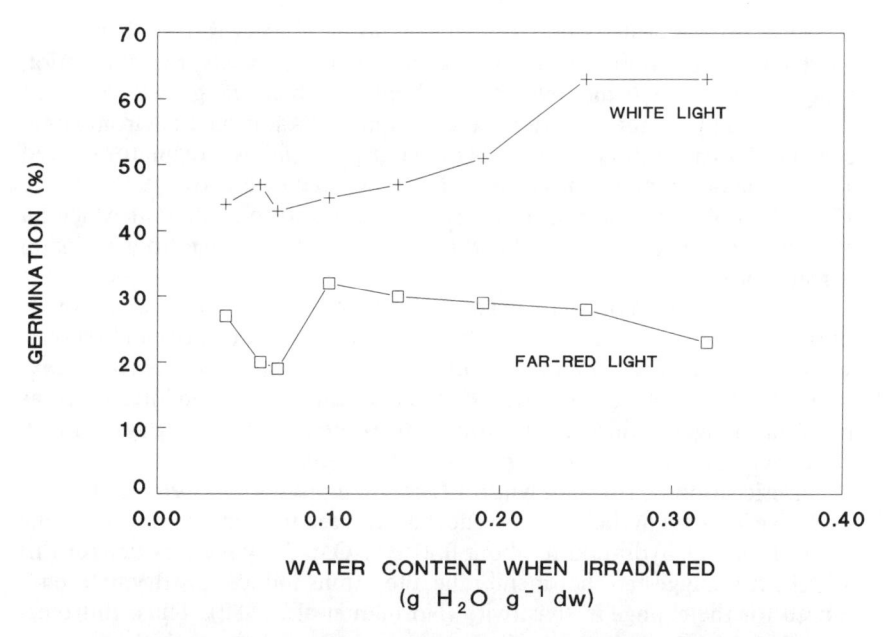

Fig. 4-8. Germination of lettuce after seeds of varying moisture contents were exposed to white light for 2 wk, followed by far-red light for 40 min. Data from Vertucci et al. (1987).

PROCESSES AFFECTED BY SEED MOISTURE CONTENT

Aging

Under optimal conditions, seed vigor deteriorates with storage time, but the cause of aging in seeds has remained elusive. Evidence of discrete hydrational regions in seeds allows us to speculate as to the mechanism of seed aging. We have described how the nature of reactions change with moisture region; thus, it is likely that the mechanisms of aging change with hydration level. Oxidative reactions are probably involved in seed deterioration (Wilson & McDonald, 1986; Flood & Sinclair, 1981). It has been suggested that autooxidations are largely responsible for aging of dry seeds (Rockland, 1969; Troller & Christian, 1978; Harrington, 1972). Since these types of reactions require large energy inputs (Priestley et al., 1985b; Boveris et al., 1983; Vertucci & Leopold, 1987b), it is unlikely that they are involved in deteriorative processes when moisture levels are in region 1, unless the reactions are radiation-induced. Enzymatic oxidations, such as those of lipoxidase, become facilitated within region 2 and may contribute to deteriorative reactions observed at this moisture level (Priestley, 1985b; Vertucci & Leopold, 1987b). Within the third water-binding region, mitochondrial respiration is possible, and the mechanism of aging at high-moisture levels may be due to utilization of food reserves or microbial attack.

Developmental Switches in Seeds

In many desiccated organisms, the addition of water is the only requirement for the resumption of physiological activity. In seeds, the desiccation process may serve as an environmental cue by which the growth phase of the organism is changed. Often, seeds require an additional environmental cue in order to germinate. The role of water potential as a regulator of seed growth and development has recently been reviewed (McIntyre, 1987; Mayer, 1986). It is of some interest to determine the moisture content at which an environmental cue is perceived and the appropriate message (to germinate) is transduced.

The desiccation of maturing seeds leads to a change in the direction of subsequent protein synthesis. This dramatic shift has been described in several species including bean (*Phaseolus vulgaris* L.) and castor bean (*Ricinus communis* L.) (Kermode & Bewley, 1986). These changes can be interpreted as physical changes in the organization of the protein-synthesizing system with the advent of drying (see Chapter 3 in this book).

Desiccation of ripening wheat (*Triticum aestivum* L.) seeds leads to an extensive increase in the responsiveness of the seed to gibberellin (Armstrong et al., 1982). Dehydration to about 0.25 g H_2O g^{-1} dw is necessary for this switch. It is suggested that membrane alterations induced by drying are required for the change in sensitivity (Norman et al., 1982). Thus, this reaction also appears to be a physical alteration of structure with the removal of water.

In contrast to the structural changes associated with drying, certain developmental changes may occur in dry seeds that are suggestive of chemical reactions. Lettuce seeds become insensitive to red light if stored at moisture contents in excess of 0.08 g H_2O g^{-1} dw (Vidaver & Hsiao, 1975). This change is prevented by anaerobiosis. Such a transformation in lettuce requires sufficient water to satisfy all the type 1 water-binding sites (Vertucci et al., 1987). Seeds of red rice (*Oryza sativa* L.) can enter a state of after-ripening, involving a loss of dormancy in dry storage. After-ripening will not occur at moisture contents below 0.05 g H_2O g^{-1} dw, and occurs poorly at moisture above 0.15 g H_2O g^{-1} dw (Leopold et al., 1987, unpublished data). Similarly, apple seeds, requiring a stratification period to break dormancy, become sensitive to temperature within the second water-binding region and obtain maximal sensitivity within the third (Vertucci & Leopold, 1986).

Thus, in addition to chemical reactions occurring as a function of seed moisture content, we can learn about the strictures of moisture content on developmental changes. We suggest that developmental changes occurring in stored seeds are a consequence of structural alterations of seed components (wheat and bean) or a consequence of nonmetabolic chemical reactions that occur in the dry seed (lettuce, red rice, and apple). These latter events may be limited to certain regions of the moisture contents. In each case, we suggest that the limits approximate the boundaries between water-binding regions.

CONCLUSION

The water content of a seed in storage is not simply a quantitative feature, but rather some qualitative differences exist in the potential for carrying out reactions relevant either to seed longevity or dormancy. The concept that there are three levels of physiological activity associated with three regions of water binding (Clegg, 1978) seems to apply to seeds. While the boundaries between the moisture regions are not precise, there appears to be an interesting association of separate biochemical processes. These repeating associations lead us to suggest that there are qualitative, as well as quantitative, differences in the biochemical transformations that occur in seeds at different moisture contents, and these differences reflect differences in the thermodynamic properties of water.

REFERENCES

Acker, L. 1969. Water activity and enzyme activity. Food Technol. (Chicago) 23:1257–1270.

Armstrong, C., M. Block, J.M. Chapman, H.A. Norman, and R. Angold. 1982. The induction of sensitivity to gibberellin in aleurone tissue of developing wheat grains. Planta 154:573–577.

Bartholomew, D.P., and W.E. Loomis. 1967. Carbon dioxide production by dry grain of *Zea mays*. Plant Physiol. 42:120–124.

Bartley, M.R., and B. Frankland. 1984. Phytochrome intermediates and action spectra for light perception in dry seeds. Plant Physiol. 74:601–604.

Bewley, J.D. 1979. Physiological aspects of desiccation tolerance. Annu. Rev. Plant Physiol. 30:195–238.

Boveris, A., A.I. Varsavsky, S. Goncalves da Silva, and R.A. Sanchez. 1983. Chemiluminescence of soybean seeds: Spectral analysis, temperature dependence and effect of inhibitors. Photochem. Photobiol. 38:99-104.

Brockmann, K.R., and L. Acker. 1977. Verhalten der lipoxygenase in wasserarmen Milieu. Lebensm. Wiss. Technol. 10:24-27.

Bull, H.B. 1944. Adsorption of water vapor by proteins. J. Am. Chem. Soc. 66:1499-1507.

Careri, G.A. Giansanti, E. Gratton, and S.N. Amprogetti. 1979. Lysozyme film hydration events: An IR and gravimetric study. Biopolymers 18:1187-1203.

Careri, G.E. Gratton, P.H. Yang, and J.A. Rupley. 1980. Correlation of IR spectroscopic, heat capacity, diamagnetic susceptibility and activity measurements on lysozyme powder. Nature (London) 284:572-573.

Chan, H.W.S. 1987. Autoxidation of unsaturated lipids. Academic Press, New York.

Ching, T.M. 1961. Respiration of forage seed in hermetically sealed cans. Agron. J. 53:6-8.

Clegg, J.S. 1978. Hydration-dependent metabolic transitions and cellular water in *Artemia* cysts. p. 117-153. *IN* J.H. Crowe and J.S. Clegg (ed.) Dry biological systems. Academic Press, New York.

Cowan, D.A., T.G.A. Green, and A.T. Wilson. 1979. Lichen metabolism. 1. The use of tritium labeled water in studies of anhydrobiotic metabolism in *Ramalina celastri* and *Peltrigera polydactyla*. New Phytol. 82:489-503.

Crowe, J.H., and L.M. Crowe. 1986. Stabilization of membranes in anhydrotic organisms. p. 188-209. *In* A.C. Leopold (ed.) Membranes, metabolism and dry organisms. Cornell Univ. Press, Ithaca, NY.

D'Arcy, R.L., and I.C. Watt. 1970. Analysis of sorption isotherms of non-homogeneous sorbents. Trans. Faraday Soc. 66:1236-1245.

Duckworth, R.B. 1962. Diffusion of solutes in dehydrated vegetables. Vol. 2. p. 46-49. *In* J. Hawthorne and J.M. Leitch (ed.) Recent advances in food science. Butterworths, London.

Flood, R.G., and A. Sinclair. 1981. Fatty acid analysis of aged permeable and impermeable seed of *Trifolim subteraneum*. Seed Sci. Technol. 9:475-477.

Harrington, J.F. 1972. Seed storage and longevity. p. 145-245. *In* T.T. Kozlowski (ed.) Seed biology. Vol 3. Academic Press, New York.

Hsiao, A.I., and W. Vidaver. 1971. Seed water content in relation to phytochrome mediated germination of lettuce seeds (*Lactuca sativa* var. Grand Rapids). Can. J. Bot. 49:111-115.

Jaenicke, R. 1981. Enzymes under extremes of physical conditions. Annu. Rev. Biophys. Bioeng. 10:1-67.

Justice, O.L., and L.N. Bass. 1978. Principles and practies of seed storage. USDA Agric. Handb. 506. U.S. Gov. Print. Office, Washington, DC.

Kendrick, R.E., and J.H. Russel. 1975. Photomanipulation of phytochrome in lettuce seeds. Plant Physiol. 56:332-334.

Kermode, A.R., and J.D. Bewley. 1986. Alteration of genetically regulated synthesis in seeds by desiccation. p. 59-84. *In* A.C. Leopold (ed.) Membranes, metabolism and dry organisms. Cornell Univ. Press, Ithaca, NY.

Labuza, T.P. 1980. The effect of water activity on reaction kinetics of food deterioration. Food Technol. (Chicago) 34:36-41.

Leopold, A.C., and C.W. Vertucci. 1986. Physical attributes of desiccated seeds. p. 22-34. *In* A.C. Leopold (ed.) Membranes, metabolism and dry organisms. Cornell Univ. Press, Ithaca, NY.

Lynch, R.M., and J.S. Clegg. 1986. A study of the metabolism in dry seeds of *Avena fatua* L. evaluated by incubation with ethanol-1-C^{14}. p. 50-58 *In* A.C. Leopold (ed.) Membranes, metabolism and dry organisms. Cornell Univ. Press, Ithaca, NY.

Mahama, A., and A. Silvy. 1982. Influence de la teneur en eau sur la radiosensibilite des semences d'*Hibiscus cannabius* L. I. Role des differents etats de l'eau. Environ. Exp. Bot. 22:233-242.

Mayer, A.M. 1986. How do seeds sense their environment? Some biochemical aspects of the sensing of water potential, light and temperature. Isr. J. Bot. 35:3-16.

McIntyre, G.I. 1987. The role of water in the regulation of plant development. Can. J. Bot. 65:1287-1298.

Morohashi, Y. 1978. Development of respiratory metabolism in seeds during hydration p. 225-240. *In* J.H. Crowe and J.S. Clegg (ed.) Dry biological systems. Academic Press, New York.

Nawa, Y., and T. Asahi. 1973. Relationship between the water content of pea cotyledons and mitochondrial development during the early stage of germination. Plant Cell Physiol. 14:607-610.

Norman, H.A., M. Black, and J.M. Chapman. 1982. Induction of sensitivity to gibberellic acid in aleurone tissue of developing wheat grains. II. Evidence for temperature dependent membrane transitions. Planta 154:578–586.

Oelze-Karow, H., and H. Mohr. 1972. Repression of lipoxygenase synthesis in plant tissue through a treshold mechanism. Proc. Int. Congr. Photobiol. Abstr. 6th, 1972: 162A.

Opik, H., and E.W. Simon. 1963. Water content and respiration rate of bean cotyledons. J. Exp. Bot. 14:299–310.

Potthast, K. 1978. Influence of water activity on enzymic activity in biochemical systems. p. 323–342. *In* J.H. Crowe and J.S. Clegg (ed.) Dry biological systems. Academic Press, New York.

Priestley, D.A. 1986. Seed aging. Cornell Univ. Press, Ithaca, NY.

----, B.G. Werner, A.C. Leopold, and M.B. McBride. 1985a. Organic free radical levels in seeds and pollen. Physiol. Plant. 64:88–94.

----, ----, and ----. 1985b. The susceptibility of soybean lipids to artificially enhanced atmospheric oxidation. J. Exp. Bot. 36:1653–1659.

Rockland, L.B. 1969. Water activity and storage stability. Food Technol. (Chicago) 23:1241–1251.

Rupley, J.A., E. Gratton, and G. Careri. 1983. Water and globular proteins. Trends Biochem. Sci. 8:18–22.

Schoper, P. 1977. Phytochrome control of enzymes. Annu. Rev. Plant Physiol. 28:223–252.

Troller, J.A., and J.H.B. Christian. 1978. Water activity and food. Academic Press, New York.

Vertucci, C.W., J.L. Ellenson, and A.C. Leopold. 1985. Chlorophyll fluorescence characteristics associated with hydration level in pea cotyledons. Plant Physiol. 79:248–252.

----, and A.C. Leopold. 1984. Bound water in soybean seed and its relation to respiration and imbibitional leakage. Plant Physiol. 75:114–117.

----, and ----. 1986. Physiological activities associated with hydration level in seeds. p. 35–49. *In* A.C. Leopold (ed.) Membranes, metabolism and dry organisms. Cornell Univ. Press, Ithaca, NY.

----, and ----. 1987a. Water binding in legume seeds. Plant Physiol. 85:224–231.

----, and ----. 1987b. Oxidative processes in soybean and pea seeds. Plant Physiol. 84:1038–1043.

----, F.A. Vertucci, and A.C. Leopold. 1987. Water content and the conversion of phytochrome regulation of lettuce dormancy. Plant Physiol. 84:887–890.

Vidaver, W., and A.I. Hsiao. 1975. Secondary dormancy in light-sensitive lettuce seeds incubated anaerobically or at elevated temperatures. Can. J. Bot. 53:2557–2560.

Wilson, A.T., M. Vickers, and L.R.B. Mann. 1979. Metabolism in dry pollen. A novel technique for studying anhydrobiosis. Naturwissenschaften 66:53–54.

Wilson, D.O., and M.B. McDonald. 1986. The lipid peroxidation model of seed aging. Seed Sci. Technol. 20:1–100.

Wilson, S.B., and W.D. Bonner. 1971. Studies of electron transport in dry and imbibed peanut embryos. Plant Physiol. 48:340–344.

5 Measurement of Seed Moisture[1]

D. F. Grabe

Oregon State University
Corvallis, Oregon

Moisture content is intimately associated with all aspects of physiological seed quality. Proven relationships exist between moisture content and seed maturity; optimum harvest time; longevity in storage; economies in artificial drying; injuries due to heat, frost, fumigation, insects and pathogens; mechanical damage; and seed weight. Since seed moisture and its management influence so many physiological quality factors essential to quality control, seed moisture measurement procedures appropriate to the purpose are needed in commerce and research.

The voluminous literature on determination of moisture content in a multitude of products testifies that there is no single method of moisture determination that is satisfactory for all products or situations. The same is true for moisture testing of seeds and grains. The optimum method for moisture testing depends on the chemical composition and seed structure; moisture content level; degree of accuracy and precision required; and constraints of time, technical expertise, and cost.

The ideal moisture-testing procedure would be one that is adapted to all seeds, measures moisture content from 0 to 1 g H_2O g^{-1} fresh weight (fw)[2], is accurate to within ± 0.001 g H_2O g^{-1} fw, is repeatable to within ± 0.001 g H_2O g^{-1} fw, takes <1 min to perform, required little training, and is low in cost. Unfortunately, it is impossible to combine all of these attributes into one procedure. Those methods that can measure the entire moisture range are not always quick, and those that are most accurate are not always inexpensive.

It is the purpose of this chapter to describe the major features of seed moisture testing methods, the principles on which they are based, their advantages and disadvantages, and appropriate applications in research and industry. Noteworthy earlier reviews were prepared by Guilbot et al. (1973) and Hunt and Pixton (1974).

---

[1] Oregon Agric. Exp. Stn. Tech. Paper no. 8448.
[2] To convert (g H_2O g^{-1} fw) to (%), multiply by 100. Thus, 0.10 g H_2O g^{-1} fw = 10%.

Copyright © 1989 Crop Science Society of America, 677 S. Segoe Rd., Madison, WI 53711, USA. *Seed Moisture*, CSSA Special Publication no. 14.

PROBLEMS IN DETERMINING SEED MOISTURE

If seeds were like grains of sand, there would be no problems associated with moisture determination by oven-drying methods. In the case of sand, the sand-water mixture can be heated at any temperature above 100 °C until no change in weight can be detected. Weight loss is due entirely to water loss, and there is no change in the physical condition of the sand particles. Moisture determinations would be absolutely accurate and repeatable.

Problems in moisture measurement of seeds are imposed by the chemical composition of the seed and the interactions of seed with water. Water is held in the seed with varying degrees of strength, ranging from free water to chemically bound water (Hunt & Pixton, 1974). During drying, the free water is removed quickly and easily, while the bound water is removed with difficulty. Becker and Sallans (1956) state that free water can be removed by the normal heat of vaporization, while removal of bound water requires heat in excess of this amount.

The water-binding relationships in seeds are frequently described in terms of equilibrium moisture curves (water-sorption isotherms). Seeds are hygroscopic, giving off or taking up water from the surrounding air until the moisture content of the seed and the relative humidity (RH) of the air are in equilibrium. When seed moisture content is plotted against the RH of the air, the resulting isotherm is in the shape of a sigmoid curve with three distinct regions (Hunt & Pixton, 1974). In the low humidity range (about 0-20%), the curve is concave to the RH axis. This represents water that is tightly bound to large protein and starch molecules. In the mid-range of the isotherm (about 20-60% RH), the curve is usually linear, and the water is less tightly bound. In the high humidity range, the curve is convex to the RH axis, rising sharply with each increase in RH above 80%. This water is considered to be free water, is loosely held by capillary forces, acts as a solvent, and is easily removed during drying.

Bound water has different physical properties than water by itself. It has a lower freezing point, higher boiling point, lower vapor pressure, and higher density (Shanbhag et al., 1970). Bound water does not act as a solvent for mineral salts and cannot conduct electricity, so it is not measured by conductivity-type moisture meters (Hunt, 1965). There is no sharp line of demarcation between bound and free water since not all bound water is held with equal force. A state of equilibrium exists between free and bound water (Hart & Golumbic, 1963), and the weight of bound water per unit weight of dry matter increases with increasing total moisture content until a maximum is reached (Shanbhag et al., 1970).

Several concepts have been proposed to explain the forces that hold water in seeds. These have been reviewed by Hunt and Pixton (1974), Pande (1974), Mitchell and Smith (1977), and Vertucci and Leopold (1984).

Bound water is so tightly held that often it cannot be removed by drying without removing other volatile substances as well, especially in seeds of high oil content. This has been attributed to accelerated decomposition from enzymatic activity as the sample is heating up to the oven temperature (Hunt & Pixton, 1974). Thus, loss of volatiles cannot be avoided by drying at lower temperatures, even in a vacuum oven at 50 °C.

Hart (1972) identified many volatile compounds given off by maize (*Zea mays* L.) during heating at 103 °C for 72 h. These compounds included CO_2, NH_3, amines, carbonyls, acids, and alcohols, in addition to water. These losses accounted for errors in moisture determination ranging from about 0.002 to 0.008 g H_2O g^{-1} fw when drying maize, 0.006 g H_2O g^{-1} fw in flax (*Linum usitatissimum*), and 0.005 g H_2O g^{-1} fw in soybean [*Glycine max* (L.) Merr.]. Nonaqueous losses in other grains were insignificant.

Chemical decomposition occurs at 130 °C in seeds of high oil content, with the production of CO_2 and water. The water that is released adds to the overall amount of water that is measured (Hart & Golumbic, 1963).

The actual weight loss during drying, therefore, is due to loss of water and other volatiles; addition of water by chemical decomposition; and reduction of dry matter by decomposition. Drying periods in official methods must be chosen to leave enough water in the seed to compensate for the weight of volatiles driven off.

The need for basic reference methods when developing practical moisture-testing methods was pointed out by Guilbot et al. (1973).

CLASSIFICATION OF SEED MOISTURE-TESTING METHODS

Moisture-testing methods may be classified as either *primary* or *secondary* (Hart & Golumbic, 1963). Primary methods are direct methods in which the water is removed and the amount determined quantitatively. There are three principal methods of removing and measuring water during a moisture test: (i) drying with heated air in a drying oven and calculating moisture loss as loss in weight of the original sample; (ii) removal by distillation and measuring the amount volumetrically in a graduated vessel; and (iii) extraction in a solvent such as methanol and determining the quantity titrimetrically, chromatographically, or spectrophotometrically. These methods measure water directly, do not require calibration against some other method, and are themselves used in calibrations. Widely accepted primary methods are considered basic or reference methods for establishment of secondary or practical methods. Although several basic methods have been used for this purpose, there is no general agreement among countries or organizations as to which method is best. Primary methods are generally the most accurate, but require an excessive amount of time for practical use for seed production and technology purposes.

Secondary, or indirect, methods measure some chemical or physical characteristic of seeds that is related to moisture content. These measurements are calibrated against the true moisture content determined by a primary reference method. The most common secondary methods measure electrical properties of seeds. Other methods utilize rapid drying, hygrometry, near-infrared spectroscopy, nuclear magnetic resonance, microwave spectroscopy, and chemical reactions. Regardless of the accuracy and precision of measurement of the property concerned, secondary methods can be no more accurate than the primary reference method against which they are calibrated. Although somewhat less accurate than primary methods, secondary methods are much faster and more practical for field use.

PRIMARY METHODS

Oven-Drying Methods

Procedure

In the air-oven method, a weighed quantity of seed is dried in an oven at a certain temperature for a specified time, cooled in a desiccator, reweighed, and the loss in weight calculated as moisture content. The oven method is often considered a basic method, but it is in fact rather empirical, with results depending on the time and temperature of drying. No single oven procedure can be used with the same degree of accuracy on all seed kinds. Proper test conditions must be determined for each kind by calibration with another primary method in which results are not dependent on drying temperature and time.

Equipment

Since air-oven methods are empirical in nature, it is essential that standardized equipment and procedures be employed to achieve accuracy and precision of moisture determinations. Equipment requirements are normally specified by the various organizations that publish official air-oven methods. These specifications vary somewhat in detail, but the general requirements are similar to the following:

Drying Oven. Either gravity-convection or mechanical-convection ovens may be used. Mechanical-convection types are preferred because they recover the set temperature quicker after opening the door to place samples. Recovery time should preferably be 15 min or less. The oven should be able to hold the required temperature within $\pm 1.0\,°C$.

Thermometer. The thermometer should read accurately to within $\pm 0.5\,°C$.

Grinder. The grinder should be adjustable so that the specified particle sizes can be obtained. The grinder should not expose the sample to the air or cause undue heating.

Moisture Dishes. An aluminum dish measuring 55 × 15 mm, with a tight-fitting lid, is desirable. Glass containers are also acceptable. The size should be such that the sample distribution does not exceed 0.3 g cm^{-2}.

Balance. The analytical balance should be quick weighing and accurate to ±1 mg.

Desiccator. The desiccator should be air-tight with a thick metal plate to promote rapid cooling of the moisture dishes. Suitable desiccants include phosphorus pentoxide, activated alumina, molecular sieves Type 4A, or calcium sulfate.

Effect of Procedures on Accuracy of Air-Oven Tests

The oven method seems deceptively simple, but many test variables must be controlled to ensure accurate and repeatable results. Sources of error were described by Hunt and Neustadt (1966), Hunt and Pixton (1974), Pande (1974), and Mitchell and Smith (1977), among others. The following comments regarding sources of error and their reduction are based on information provided by these authors.

Oven. Each oven should be evaluated for uniformity of heating between shelves and at various locations on a shelf. Uniformity of temperature can best be ensured by utilizing only one shelf and placing the samples in locations with uniform temperature. The bulb of the thermometer should be placed at seed level. Even thermometers should be checked occasionally for accuracy since long exposure to high temperature may cause them to become inaccurate.

Drying Temperature. Drying to constant weight is not a valid indication of true moisture content because different constant weights are attained at different temperatures (Hart et al., 1959). In many cases, however, temperature and drying period appear to have been established empirically, based on attainment of constant weight.

Drying Period. The drying period, especially at 130°C, is extremely critical. Timing of the drying period should begin when the required temperature is regained after opening the door and placing the samples. At 130°C, more than 90% of the water is driven from the seeds during recovery time (Hart & Golumbic, 1963).

It is not possible to select a single drying period that will provide the same degree of accuracy for seed at different moisture levels. Benjamin and Grabe (1988) showed that water is removed faster from high-moisture grass seeds than from low-moisture seeds. The best that can be done is to select a compromise drying period for the majority of seed samples, and these are usually air-dry.

Moisture Dish. The shape and composition of the drying dish will affect the speed of drying and accuracy of weighing. Tall narrow glass bottles are less satisfactory than flat aluminum dishes because the seed depth is excessive, increasing the drying time. With heavy glass bottles, the majority of the weighing error is reflected in the weight of the bottle rather than in the light weight of the seed. Some plastic weighing bottles lose weight during heating.

Sample Size. For a particular size of drying dish and drying period, larger samples may give lower moisture values.

Grinding. Grinding of large seeds shortens the drying period. Drying periods in official methods are calibrated for a particular particle size, usually 18 or 20 mesh. Particles larger than prescribed will not dry completely during the established drying period. Exposure of the seed to air during grinding, and excessive heating in some grinders, may change the moisture content during grinding. Williams and Sigurdson (1978) tested four grinders and found that all four lost moisture when the grain moisture content was about 0.10 g H_2O g^{-1} fw.

Room Humidity. At a drying temperature of 103 °C, moisture values will be low if the RH of the air in the room is above 90%. Humidity has no practical effect at 130 °C, although it has been demonstrated that seed will absorb moisture if a pan of water is placed in the 130 °C oven (Hunt & Neustadt, 1966).

Desiccant. Weighing errors are introduced if the dried sample in an open dish is removed from the oven and weighed while still hot. The two principal causes are absorption of moisture from the air and a buoyancy effect due to warm air enclosed in the sample (Pande, 1974). To prevent these errors, the lids are placed on the drying dishes while still in the oven and the dishes are cooled for at least 45 min in a desiccator before weighing. Silica gel should not be used as a desiccant as it will give up moisture to dried seeds (Hunt & Neustadt, 1966). Calcium chloride is also not suitable.

Development of Standard Air-Oven Methods

Because of the empirical nature of air-oven methods and the many variables that must be controlled to obtain precise results, several national and international organizations have developed standard air-oven methods for testing moisture content. Most of these organizations deal primarily with cereals, cereal products, and oil seeds. Extensive research has gone into development of official methods because of the commercial need for standardization. Official air-oven methods for cereals were developed by calibration against basic reference methods. The vacuum-phosphorus pentoxide method was used for calibrating the oven methods followed by the International Association for Cereal Chemistry (ICC, 1976), International Organization for Standardization (ISO, 1985), and the International Seed Testing Association (ISTA, 1985). Drying to constant weight in a vacuum oven at 98 to 100 °C was the reference method for the air-oven procedures of the American Association of Cereal Chemists (AACC, 1979), Association of Official Analytical Chemists (AOAC, 1984), and the Official Grain Standards of the USDA (USDA, 1959). Subsequent to development of official oven methods for cereals, they have been found to give results in good agreement with the Karl Fischer basic reference method.

Unfortunately, there are few research reports detailing accurate oven methods for seeds other than cereals and oil seeds. The notable exception is the work by Hart et al. (1959) who published an extensive list of oven methods for whole seeds that were derived by comparison with the Karl Fischer method. This list includes 31 grains, grasses, legumes, oil seeds, and vegetable seeds. Most of these methods were subsequently adopted by the American Society of Agricultural Engineers (ASAE, 1987). Hart and Golumbic (1966) also published methods for 16 additional kinds based on the Karl Fischer method.

For each seed kind, an almost infinite number of time and temperature combinations will give results equivalent to those of basic reference methods. In selecting desirable combinations for a large number of species, it is more practical to limit the number of temperatures (and ovens) required and to determine the drying periods required at the designated temperatures. The temperatures most commonly used are 130 °C for high-temperature methods, and 103 or 105 °C for low-temperature methods.

A compilation of air-oven methods that have been proven against basic reference methods is presented in Table 5-1.

Air-Oven Methods of Seed-Testing Organizations

Oven methods in the International Rules for Seed Testing (ISTA, 1985) underwent several changes between the first edition in 1931 and the present edition of 1985 (Grabe, 1987). Between 1931 and 1953, whole seeds of all species were dried 5 h at 103 °C (ISTA, 1931). Several intermediate revisions were made until, in the 1985 Rules, there is a high-temperature method at 130 °C and a low-temperature method at 103 °C. Drying times at 130 °C are 4 h for maize, 2 h for other cereals, and 1 h for all other kinds. Drying time at 103 °C is 17 h. Larger seeds are ground and smaller seeds are left whole. To gain a perspective on the effect of these gradual procedural changes on the apparent moisture content of seeds, samples of wheat (*Triticum aestivum* L.) were tested by the 1931 to 1953 and 1985 methods (Grabe, 1987). The apparent moisture contents by the 1985 method were about 0.03 g H_2O g^{-1} fw higher than by the earlier method. This poses a disquieting question: Have changes in moisture testing methods over the years affected the interpretation of older research results and current recommendations regarding proper seed moisture content for the maintenance of seed quality?

The ISTA methods for some species, especially cereals, are based on basic reference methods and are acceptable in their present form. Methods for most of the other species, however, appear to have been developed empirically without comparison with basic reference methods. For those species, repeatability of results is good, but their accuracy is mostly unknown and sometimes poor (Grabe, 1984).

Numerous national and regional seed-testing organizations follow the ISTA Rules for conducting seed moisture tests.

The Association of Official Seed Analysts (AOSA, 1981) and the Society of Commercial Seed Technologists have no standard methods for testing seed moisture. Member laboratories of these organizations use many different combinations of temperature and drying period, and would obtain widely

Table 5-1. Air-oven methods for seed moisture testing reported to give results in agreement with basic reference methods.

Species	Common name	Sample preparation	Sample size (g)	Temperature (°C)	Drying period (h min)	Reference method	Reference
		Agricultural seeds					
Agrostis tenuis Sibthorp	Colonial bentgrass	Whole	10	130	1 0	Karl Fischer	Hart et al., 1959
Andropogon ischaemum	Yellow bluestem	Whole	1	100	1 0	Karl Fischer	Hart et al., 1959
Avena sativa L.	Oat	Whole	10	130	23 0	Karl Fischer	Hart et al., 1959
Avena sativa L.	Oat	Ground	5	130	2 0	Vacuum-P_2O_5	Int. Assoc. Cereal Chem., 1976
Brassica napus L. var. *annua* Koch	Annual rape	Whole	10	130	4 0	Karl Fischer	Hart et al., 1959
Bromus inermis Leyss.	Smooth brome	Whole	4	130	0 50	Karl Fischer	Hart et al., 1959
Carthamus tinctorius L.	Safflower	Whole	3	130	1 30	Karl Fischer	Hart et al., 1959
Dactylis glomerata L.	Orchardgrass	Whole	3	130	3 0	Karl Fischer	Benjamin & Grabe, 1988
Festuca arundinacea Schreb.	Tall fescue	Whole	3	130	3 0	Karl Fischer	Benjamin & Grabe, 1988
Festuca rubra L.	Res fescue	Whole	3	130	3 0	Karl Fischer	Benjamin & Grabe, 1988
Glycine max (L.) Merr.	Soybean	Ground	2–3	130	1 0	Vacuum-100°C	Am. Assoc. Cereal Chem., 1979
Helianthus annuus L.	Sunflower	Whole	10	130	1 0	Karl Fischer	Hart et al., 1959
Hordeum vulgare L.	Barley	Whole	5	130	20 0	Karl Fischer	Hart et al., 1959
Hordeum vulgare L.	Barley	Ground	5	130	2 0	Vacuum-P_2O_5	Int. Assoc. Cereal Chem., 1976
Lens culinaris Medic.	Lentil	Ground	2–3	130	1 0	Vacuum-100°C	Am. Assoc. Cereal Chem., 1979
Lespedeza stipulacea Maxim.	Korean lespedeza	Whole	12	100	2 30	Karl Fischer	Hart & Golumbic, 1966
Linum usitatissimum L.	Flax	Whole	10	103	4 0	Karl Fischer	Hart et al., 1959
Lolium perenne L.	Perennial ryegrass	Whole	5	130	3 0	Karl Fischer	Hart et al., 1959
Medicago sativa L.	Alfalfa	Whole	10	130	2 30	Karl Fischer	Hart et al., 1959
Oryza sativa L.	Rice	Ground	5	130	2 0	Vacuum-P_2O_5	Int. Assoc. Cereal Chem., 1976
Phaseolus vulgaris L.	Field bean	Whole	15	103	72 0	Vacuum-100°C	Am. Assoc. Cereal Chem., 1979
Phleum pratense L.	Timothy	Whole	10	130	1 40	Karl Fischer	Hart et al., 1959
Pisum sativum var. *arvense* (L.) Poir.	Field pea	Ground	2–3	130	1 0	Vacuum-100°C	Am. Assoc. Cereal Chem., 1979
Poa pratensis L.	Kentucky bluegrass	Whole	3	130	3 0	Karl Fischer	Benjamin & Grabe, 1988
Secale cereale L.	Rye	Whole	10	130	16 0	Karl Fischer	Hart et al., 1959
Secale cereale L.	Rye	Ground	5	130	2 0	Vacuum-P_2O_5	Int. Assoc. Cereal Chem., 1976
Setaria italica (L.) Beauv.	Foxtail millet	Ground	5	130	2 0	Vacuum-P_2O_5	Int. Assoc. Cereal Chem., 1976
Sorghum vulgare Pers.	Sorghum	Whole	10	130	18 0	Karl Fischer	Hart et al., 1959
Sorghum vulgare Pers.	Sorghum	Ground	2–3	130	1 0	Vacuum-100°C	Am. Assoc. Cereal Chem., 1979
Trifolium hybridum L.	Alsike clover	Whole	10	130	2 30	Karl Fischer	Hart et al., 1959
Trifolium incarnatum L.	Crimson clover	Whole	10	130	2 30	Karl Fischer	Hart et al., 1959
Trifolium repens L.	White clover	Whole	10	130	2 30	Karl Fischer	Hart et al., 1959
Triticum aestivum L.	Wheat	Whole	10	130	19 0	Karl Fischer	Hart et al., 1959
Triticum aestivum L.	Wheat	Ground	5	130	2 0	Vacuum-P_2O_5	Int. Assoc. Cereal Chem., 1976
Zea mays L.	Maize	Whole	15	103	72 0	Vacuum-100°C	Am. Assoc. Cereal Chem., 1979
Zea mays L.	Maize	Ground	8	130	4 0	Vacuum-P_2O_5	Int. Assoc. Cereal Chem., 1976

MEASUREMENT OF SEED MOISTURE

Vegetable seeds

Species	Common name	Form			Method	Reference
Alium cepa L.	Onion	Whole	10	130	0 50	Karl Fischer	Hart et al., 1959
Apium graveolens var. dulce L.	Celery	Whole	10	93	4 0	Karl Fischer	Hart & Golumbic, 1966
Beta vulgaris L.	Beet	Whole	5	110	2 30	Karl Fischer	Hart & Golumbic, 1966
Beta vulgaris var. cicla L.	Swiss chard	Whole	5	100	2 45	Karl Fischer	Hart & Golumbic, 1966
Brassica juncea (L.) Coss.	India mustard	Whole	10	130	4 0	Karl Fischer	Hart et al., 1959
Brassica oleracea var. acephala DC.	Collards	Whole	10	130	4 0	Karl Fischer	Hart et al., 1959
Brassica oleracea var. acephala DC.	Kale	Whole	10	130	4 0	Karl Fischer	Hart et al., 1959
Brassica oleracea var. capitata L.	Cabbage	Whole	10	130	4 0	Karl Fischer	Hart et al., 1959
Brassica rapa L.	Turnip	Whole	10	130	4 0	Karl Fischer	Hart et al., 1959
Cucumis sativus L.	Cucumber	Whole	10	130	1 40	Karl Fischer	Hart & Golumbic, 1966
Daucus carota L.	Carrot	Whole	10	100	1 40	Karl Fischer	Hart et al., 1959
Hibiscus esculentus L.	Okra	Whole	10	130	3 0	Karl Fischer	Hart & Golumbic, 1959
Lactuca sativa L.	Lettuce	Whole	10	120	1 30	Karl Fischer	Hart & Golumbic, 1959
Pastinaca sativa L.	Parsnip	Whole	5	100	2 0	Karl Fischer	Hart & Golumbic, 1959
Petroselinum crispum (Mill.) Nym.	Parsley	Whole	10	100	2 0	Karl Fischer	Hart et al., 1959
Raphanus sativus L.	Radish	Whole	10	130	1 10	Karl Fischer	Hart et al., 1959
Spinacea oleracea L.	Spinach	Whole	10	130	2 35	Karl Fischer	Hart & Golumbic, 1966

Tree seeds

Species	Common name	Form			Method	Reference
Abies grandis (Dougl.) Lindl.	Grand fir	Whole	6	100	2 0	Karl Fischer	Hart & Golumbic, 1966
Abies procera Rehd.	Noble fir	Whole	6	100	2 0	Karl Fischer	Hart & Golumbic, 1966
Abies procera Rehd.	Noble fir	Whole	10	105	1 0	Toluene	Buszewicz, 1962
Fraxinus pennsylvanica Marsh.	Green ash	Whole	--	130	4 0	Toluene	Bonner, 1972
Liquidambar styraciflua L.	Sweetgum	Whole	--	130	4 0	Toluene	Bonner, 1972
Liriodendron tulipifera L.	Tuliptree	Whole	5–8	105	15 0	Toluene	Bonner, 1979
Picea engelmannii Parry	Engelmann spruce	Whole	10	130	1 45	Karl Fischer	Hart & Golumbic, 1966
Pinus elliottii Engelm.	Slash pine	Whole	5–8	105	16 0	Toluene	Bonner, 1979
Pinus lambertiana Dougl.	Sugar pine	Whole	7	110	1 50	Karl Fischer	Hart & Golumbic, 1966
Pinus palustris Mill.	Longleaf pine	Whole	5–8	105	16 0	Toluene	Bonner, 1979
Pinus ponderosa Laws	Ponderosa pine	Whole	10	100	2 45	Karl Fischer	Hart & Golumbic, 1966
Pinus strobus L.	Eastern white pine	Whole	5–8	105	16 0	Toluene	Bonner, 1979
Pinus taeda L.	Loblolly pine	Whole	5–8	105	16 0	Toluene	Bonner, 1979
Platanus occidentalis L.	American sycamore	Whole	--	130	4 0	Toluene	Bonner, 1973
Prunus serotina Ehrh.	Black cherry	Whole	5–8	105	15 0	Toluene	Bonner, 1979
Pseudotsuga menziesii (Mirbel) Franco	Douglas fir	Whole	7	100	1 12	Karl Fischer	Hort & Golumbic, 1966
Quercus muehlenbergii Engelm.	Chinkapin oak	Cut	10	105	8 0	Toluene	Bonner, 1974
Quercus nigra L.	Water oak	Cut	10	105	9 0	Toluene	Bonner, 1974
Quercus shumardii Buckl.	Shumard oak	Cut	25	105	9 0	Toluene	Bonner, 1974

disparate results if the same sample were tested by those laboratories. Moisture content is normally not a factor in the trading of seeds, however. Discrepancies in results generally go unnoticed because it is not common for the same seed lot to be tested for moisture content by different laboratories.

Additional Considerations

Two-Stage Drying and Grinding. The official methods of several organizations provide for grinding large seeds to shorten the drying period. For example, grinding shortens the drying period of wheat from 19 to 2 h (Hart et al., 1959). But grinding, if not carefully done, may contribute to errors of determination by exposure of the ground material to the humidity of the air, decomposition by heat, and loss of material during handling.

It is not possible to grind high-moisture seeds to the specified particle size with most grinders. When the moisture content is above 0.17 g H_2O g^{-1} fw {0.10 g H_2O g^{-1} fw for soybean and 0.13 g H_2O g^{-1} fw for rice [*Oryza sativa* (L.)]}, the unground seeds should be dried to that level before grinding (Hunt & Neustadt, 1966). The loss in weight during predrying is added to that lost during the oven-drying period.

Seeds with more than 0.25 g oil g^{-1} fw oil content may become gummy during grinding and cannot be ground satisfactorily. If seeds contain oil with an iodine number higher than 150, the oil in ground material will oxidize readily when heated, causing a gain in weight of dry matter (Zeleny, 1953).

The additional weighing and handling during the two-stage drying procedure and grinding can lead to additional errors as well as greater inconvenience.

For these reasons, there are distinct advantages in determining moisture content on whole seeds when time is not a constraint.

Calculation of Moisture Content on Wet (Fresh) or Dry Weight Basis. In the grain and seed industries, moisture content is determined on a wet-weight basis; that is, loss in weight during drying is divided by the original (wet or fresh) weight (fw). Calculation on a fw basis can present difficulties in certain research situations because changes in seed weight are not directly proportional to changes in moisture content. In some cases, research workers prefer to calculate moisture content on a dry weight (dw) basis for a clearer presentation of their data. In reporting research data, it is essential to state whether the moisture contents are on a fw or dw basis.

Equations and a scale for converting moisture percentages between the fw and dw basis are contained in the ASAE Standards (1987). The conversion equations are:

$$\% \text{ moisture, dw basis} = \frac{100 \times \% \text{ moisture, fw basis}}{100 - \% \text{ moisture, fw basis}} \quad [1]$$

$$\% \text{ moisture, fw basis} = \frac{100 \times \% \text{ moisture, dw basis}}{100 + \% \text{ moisture, dw basis}}. \quad [2]$$

(Also see Fig. 5-1.)

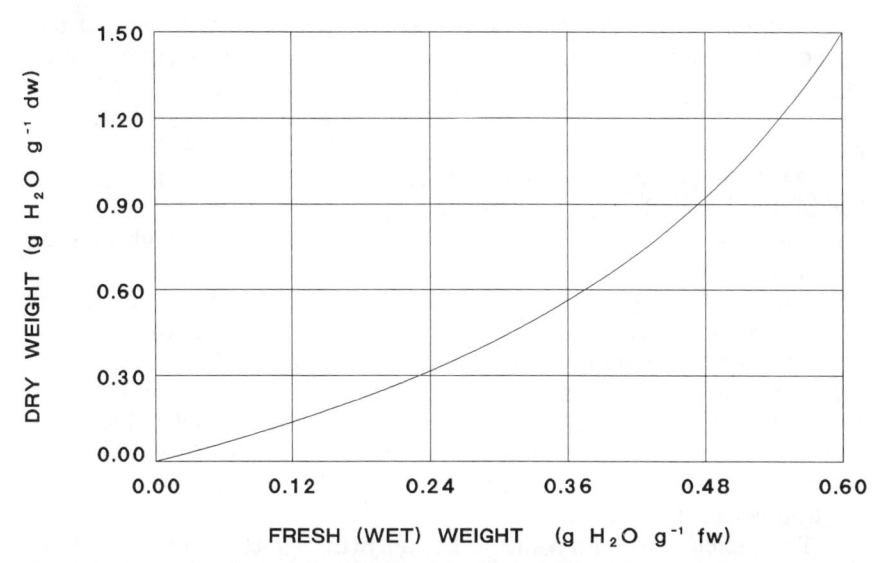

Fig. 5-1. Comparison between fresh (wet) weight (fw) and dry weight (dw) reporting of sample moisture content (P.C. Stanwood, 1987, unpublished data).

Other Types of Drying Methods

Brabender Oven. In the Brabender oven, a balance is incorporated for weighing samples without removing them from the oven, eliminating the need for a separate balance, desiccator, and weighing dishes. This type of equipment is a time-saver when used in research to develop drying curves at different temperatures (Pande, 1974).

Vacuum Oven. In vacuum oven methods, seeds are dried at temperatures lower than 100 °C in a partial vacuum until constant weight is attained. Nonaqueous volatiles may still be lost, however, because they are released at lower temperatures under vacuum (Hart & Golumbic, 1962). Constant weight of cereal seeds is attained after about 5 h at 98 to 100 °C at a pressure of 25 mm mercury or less (AACC, 1979).

A variation of the vacuum-oven method is the use of drying tubes in which phosphorus pentoxide is placed to absorb the moisture released by the seed. Constant weight is attained after about 150 h at 50 °C (ICC, 1976).

Drying to constant weight in a vacuum oven has served as a basic reference method for several official air-oven methods for cereals (Table 5-1).

Distillation Methods

Toluene Distillation

The apparatus for toluene distillation consists of a flask connected to a Leibig condenser by a distilling trap of the Bidwell and Sterling type with 0.1-mL graduations (Bidwell & Sterling, 1925). The weighed sample is placed in the flask with toluene, brought to a boil, and the water distilled until no more is given off. The water is collected in the trap where the amount is determined volumetrically. The weight of the water is then calculated by knowing the weight of water at its temperature when measured (ISTA, 1966). The test is usually completed in 1 to 5 h, although Hart and Golumbic obtained more water after 21 h than after 5 h (Hart & Golumbic, 1962).

Distillation methods were developed to overcome some of the shortcomings of the oven method, particularly the weight loss from removal of volatiles other than water. The method also prevents oxidation of the sample because of less access to O_2.

The distillation procedure has several shortcomings, including some of those inherent in the oven method. The amount of water distilled is dependent on the boiling point (bp) of the distilling liquid. Hart and Golumbic (1962) found that more water is distilled in a xylene-toluene mixture with a bp of 130 °C than in toluene (bp = 111 °C) or benzene (bp = 80 °C). There are also problems in making accurate volumetric readings, and some water may collect on the sides of the apparatus rather than flowing into the bottom of the trap where it is measured.

The toluene distillation method is an official method of the AACC (1979). It was also included in the ISTA Rules for many years but was discontinued in 1985 (ISTA, 1985).

Buszewicz (1962) and Bonner (1972, 1974, 1979) developed oven methods for several kinds of tree seeds based on comparison of results with the toluene distillation method (Table 5-1).

Brown-Duvel

The Brown-Duvel distillation method, developed in 1907, was for many years an official method for grain inspection in the USA. A 100-g sample of whole seed is heated in a flask with 150-mL nonvolatile oil to a specified cutoff temperature (180 °C for wheat) that varied with different seeds. The water is distilled into a graduated cylinder in about 1 h (Hlynka & Robinson, 1954). The method can be considered a direct, or primary, method since water is removed and measured, but it is also a practical method in that it takes a relatively short time and the procedures are not complicated. Commercial equipment contains up to six heating units and could be used in country elevators. The procedure is highly empirical, with accuracy and precision depending on precise temperature control. High temperatures are used, resulting in water gain from decomposition of organic matter. The method needs to be calibrated against a basic reference method. The Brown-Duvel method for cereals has been largely replaced by electric meters.

Solvent Extraction Methods

Solvent extraction methods appear to be the most accurate of the primary moisture-testing procedures. These methods are superior to oven methods because results are not dependent on temperature and drying time, and nonaqueous volatiles are not measured as water. In performing the test, water is removed from ground seed by methanol extraction and the water content of the extract determined precisely by Karl Fischer titration, infrared spectrophotometry, or gas chromatography. However, the accuracy of the test is more dependent on the completeness of water extraction than on the accuracy of measurement of water content of the extract.

Extraction of Water

Methanol is superior to methoxy ethanol, ethylene glycol, or dioxane for extracting water from seeds (Hart & Neustadt, 1957).

Wiese et al. (1965) obtained complete extraction of water from ground cereal seed by soaking 4 g of ground seed in 50-mL of methanol in a 50-mL Erlenmeyer flask at room temperature. Extraction times varied from 1 to 5 d. By using the same method, complete extraction of grass seed occurred within 1 d (Benjamin & Grabe, 1988).

Several procedures have been developed for shortening the extraction time to put the test on a more practical basis. Hart and Neustadt (1957) ground the dry cereal seed 3 min at 15 000 rpm in a modified Stein mill, followed by a 5-min wet grind in methanol at 64.5 °C. Kostyrko and Plebanski (1965) developed a grinder-extractor in which they ground dry seed at 13 000 rpm and extracted the water by grinding for 3 min at 63 °C in methanol. Their grinder gave 87% of the particles smaller than 0.5 mm. Jones and Brickenkamp (1981) and Robertson and Windham (1983) extracted by milling seed and methanol in a ball mill for 15 and 30 min. The process of milling heated the methanol-water mixture to just below the boiling point of methanol. Paynter and Hurburgh (1983) extracted maize seed by grinding with methanol in a household blender for 1 min at 12 500 rpm and 1 min at 17 000 rpm, followed by gentle stirring for 15 min. The extraction methods described were all evaluated for completeness of extraction. Completeness and speed of extraction vary with fineness of grind, temperature, agitation, and kind of seed.

Karl Fischer Method

The Karl Fischer method for determining moisture content is based on titration of the methanol-water mixture with a solution containing methanol, iodine, sulfur dioxide, and pyridine, the so-called Karl Fischer reagent, which is named after its developer (Fischer, 1935). As long as any water is present, the iodine is reduced to colorless hydrogen iodide. The titration end point is the first appearance of free iodine, determined either electrometrically, or visually by the appearance of brown iodine color. Pyridine in the original Karl Fischer reagent is a toxic and unpleasant-smelling compound. Modified reagents have recently been formulated in which the pyridine has been replaced by amines to eliminate the unpleasant smell of the old Karl Fischer solution (Scholz, 1981). The Karl Fischer reagent is specific for water, with no interfering substances in the seed extracts that react with the solution (Hart & Neustadt, 1957). The procedure is, therefore, probably the most basic and accurate method available. Thorough discussions of the Karl Fischer titrimetric procedures have been published by Mitchell and Smith (1980) and Pande (1974).

Two types of automatic Karl Fischer titrators are available commercially. Both employ electrometric end-point detection. The type most suitable for seed moisture testing uses a motorized buret to dispense the Karl Fischer reagent. It is capable of measuring moisture contents from 0 to 1 g H_2O g^{-1} fw and can complete the titration in <10 min. The other type is based on coulometric generation of iodine, eliminating the buret. It is more appropriate for measuring trace amounts of water in extremely small samples.

Methods of performing Karl Fischer moisture determinations on seeds with automatic Karl Fischer titrators were described by Jones and Brickenkamp (1981), Paynter and Hurburgh (1983), Robertson and Windham (1983), and Benjamin and Grabe (1988).

Near-Infrared Spectrophotometry

Hart et al. (1962) measured moisture content of methanol extracts of seeds by near-infrared (NIR) transmission spectroscopy, using the absorbance band of water at 1.93μ. Water content was found by reference to a calibration curve prepared from methanol samples containing known amounts of water. The results agreed with those of the Karl Fischer method on the same samples.

Gas Chromatography

Wiese et al. (1965) used gas chromatography to measure moisture content of methanol extracts of grains and obtained results in agreement with those of the Karl Fischer method. It thus appears that gas chromatography and NIR transmission spectroscopy may be used alternatively to the Karl Fischer procedure as a primary standard reference method. These methods, however, have not received much attention by seed researchers and are not in routine use.

SECONDARY METHODS

Electrical Methods

Electric moisture meters have achieved great popularity for testing moisture of seeds and grains. Their speed, convenience, ease of operation, and relative accuracy more than compensate for some loss of accuracy compared to primary methods. Electric meters are widely used for cereals, oilseeds, edible legumes, and other free-flowing seeds. However, calibrations are not available for testing many grass, vegetable, flower, and tree seeds.

Measurement of moisture with electric meters is based on the principle that certain electrical properties of seeds are related to their moisture content. These meters generally measure the conductivity or capacitance of seeds, although at least one meter incorporates principles of both. Pande (1975) has provided detailed information on principles of operation and electrical circuitry of conductance and capacitance moisture meters. Multon (1979) compared the characteristics of a number of electric meters with acceptable accuracy.

Conductance Meters

In conductance (resistance) meters, a direct current is imposed between two electrodes. Seed is placed between the electrodes, usually under pressure, and the amount of current is measured (Hunt, 1965). The conductivity increases with an increase in moisture content.

According to Pande (1975), the effectiveness of conductivity meters is limited to the hygroscopic range of seed moisture between about 30 and 90% RH (0.07 and 0.23 g H_2O g^{-1} fw moisture content). At moisture levels below about 0.07 g H_2O g^{-1} fw, moisture is tightly bound and does not act as a solvent for mineral salts. Thus, it cannot conduct electricity and is not measured by conductance-type meters. Since only free water is measured, calibrations are obtained by adding a constant value for bound water. However, it is probable that the ratio between free and bound water does not remain constant, and errors are introduced as changes in this ratio occur. At moisture levels above about 0.23 g H_2O g^{-1} fw, conductivity is so great that it increases little with further increases in moisture content (Hunt & Pixton, 1974).

Accuracy of conductance meters depends on uniform distribution of moisture within the seed because these meters tend to measure the conductivity on the seed surface. Low readings will be obtained from seeds that have been recently artificially dried since the surface will be drier than the interior. Recently moistened seeds will give high readings because the surface will be wetter than the interior. Moisture should also be equally distributed throughout the sample since conductivity meters measure the path of least resistance, and a single wet spot will cause erroneous readings (Pande, 1975).

Moisture content readings must be corrected for temperature since a difference of 5 °C will result in about 0.01 g H_2O g^{-1} fw difference in apparent moisture content (Pande, 1975).

It is necessary to apply a uniform pressure on the seeds to obtain dependable results, since conductivity of the sample increases when the pressure is increased.

Conductance meters are relatively easy to keep in proper adjustment and in alignment with other meters. At one time, conductance meters were used for official grain grading. Their use has declined because of their limited moisture range and inaccuracy when monitoring artificial drying. They have been largely replaced by capacitance-type meters.

Capacitance Meters

In capacitance (dielectric) meters, samples of known weight are exposed to a high-frequency voltage of 1 to 20 MHz in the meter test cell. Some of the waves are absorbed by the H atoms of the water molecules. The strength of this absorption, or the ability to store an electrical charge, is known as the *dielectric constant* (Nelson, 1981). Measurement of the dielectric constant is translated into percentage moisture by charts, direct-reading potentiometers, or microprocessors (Hurburgh et al., 1985).

Measurement of the dielectric constant is related to the seed's moisture content. Pure water has a dielectric constant of about 4, close to that of the starch or protein molecule with which it is associated (Hunt, 1965). Any variation in the amount of bound water will also influence moisture determination by capacitance moisture meters.

The dielectric constant at any given frequency increases in value with moisture content and temperature. The relationship between the dielectric and moisture content is nonlinear throughout the range of moisture levels. Temperature corrections are made automatically or by means of temperature correction charts. Packing density of the seed in the cell can affect the capacitance measurement. Proper instrument design can minimize packing-density variation among free-flowing seed samples, but cannot accommodate the large variation in packing of fluffy seeds such as the chaffy grasses. Nelson (1981) has reviewed the factors influencing the dielectric properties of cereal grains.

Capacitance moisture meters are more reliable than conductance meters when monitoring the progress of artificial drying because their accuracy is less affected by the drier outer layers of the seed. Capacitance meters are capable of measuring moisture up to about 0.4 g H_2O g^{-1} fw, but they are most accurate between about 0.06 and 0.25 g H_2O g^{-1} fw. This is unfortunate since many seed crops are harvested at moisture contents above this level.

Many models of capacitance meters are commercially available, ranging from inexpensive farm-type meters to relatively expensive trade-type meters. Hurburgh et al. (1986) found that the coefficient of variability of farm-type meters was higher than that of trade-type meters, and the variability of all meters tested increased with moisture content. Compared to conductance meters, it is more difficult to obtain repeatable results with capacitance meters, meter-to-meter variation is greater, and it is more difficult to keep them accurately calibrated.

Meters must be calibrated for each kind of seed, with calibrations based on many samples from different years, areas, and varieties. However, the electrical properties of any particular sample may vary from the overall average due to year-to-year and area-to-area differences. The electrical properties may change from day to day on some samples although the moisture values determined by the oven method have not changed (Hunt & Neustadt, 1966).

Rapid Drying Methods

Direct Weighing Balance

Variations of the oven-drying method have been developed to shorten the test period for practical use by seed growers and seed companies. These generally involve the incorporation of a heat source above the pan of a direct-reading balance. The heating unit may be an electric heating element or an infrared bulb. Although the seed sample is exposed to the air, the amount of moisture absorbed from the air is apparently negligible at the high temperatures employed. This method is especially useful for testing chaffy grasses. Attainment of constant weight is often taken as the criterion for determining moisture content, but the same sources of error are present as in oven methods. For greatest accuracy, methods employing direct-reading balances should be calibrated against a basic reference method.

Engine Exhaust Dryers

Portable dryers have been developed that connect to an engine exhaust for a heat source so they can be used in the field. Seeds are dried to constant weight and weighed on a portable scale. Dexter (1947) described a meter that attached to a tractor exhaust, while Klein and Harmond (1971) developed a version that attached to an automobile exhaust.

Microwave Oven

A currently manufactured moisture analyzer consists of a microwave drying oven, an electronic balance, and a microprocessor digital computer. The material is dried in a few minutes by exposure to microwave radiation. By selectively heating and removing the water in a short time, it is claimed that there is less loss of volatile components than in conventional ovens. This equipment is expensive, however, and the ordinary household microwave oven can be adapted to moisture testing, with drying to constant weight and external weighing (Backer & Walz, 1985).

Hygrometric Methods

Hygrometric methods utilize the hygroscopic nature of seeds to measure their moisture content. When seeds are placed in a sealed container, the RH of the enclosed air establishes an equilibrium with the moisture content of the seed. The RH of the air is determined and converted into seed moisture content by comparison with the equilibrium moisture curve for that seed. The method is nondestructive to seeds. Successful application of this technique is dependent on an accurate method of measuring RH, accurate equilibrium moisture curves for each kind of seed to be tested, and temperature control.

Relative humidity may be determined by several methods in moisture-testing applications. A commercially available instrument incorporating a humidity sensor in a sealed chamber is said to have an accuracy of ±2.0% RH. Output from the device will either provide a direct reading of RH or a constant digital printout on a data logger.

A container-type hygrometer is available in which a hygrometer with a bimetallic element is placed on top of a container holding the seed. After an equilibration period, the RH is indicated on a direct-reading dial.

Dexter (1948) described the use of a mixture of ferric ammonium sulfate and potassium ferrocyanide as a colorimetric indicator of RH. The mixture is blue at low humidities and turns red at high humidities. It is said to be accurate within 0.01 g H_2O g^{-1} fw in the moisture range of 0.08 to 0.12 g H_2O g^{-1} fw for wheat. In another application of this principle, paper strips impregnated with the proper salt solutions for measuring the entire RH range are available commercially and provide an inexpensive method for making crude measurements.

Hygrometric moisture meters may have special uses where other types of moisture meters are inadequate. One potential application being investigated is the measurement of moisture content of pelleted seed. This is difficult to measure accurately in an electric meter since the seed and pelleting material may be at different moisture levels when in equilibrium with the air.

Accurate equilibrium moisture content data are best determined by holding seed samples over a series of saturated salt solutions to provide the required RH (Winston & Bates, 1960). Numerous tables of hygroscopic equilibria of seeds have been compiled (Roberts, 1972; Justice & Bass, 1978; ASAE, 1987), but none of these are complete for all seed kinds.

The relationship between seed moisture content and the RH of the surrounding air is not absolute, but depends on several factors. At a given RH, the equilibrium moisture content will be higher during desorption and lower during adsorption. The moisture content difference between adsorption and desorption decreases with repeated wetting and drying cycles. Other variables include varietal differences, seed maturity and deterioration, oil and protein content, and accuracy and repeatability of the RH and moisture-measuring procedures.

Hygrometric moisture meters have additional practical limitations. They are limited to a range of moisture contents corresponding to RH below 100%—about 0.25 g H_2O g^{-1} fw in seeds. The method is not applicable to seeds with higher moisture contents because the RH remains at 100%. Accuracy is greatest in the linear region of the sorption isotherm. Accuracy is less in the upper and lower regions because a percent change in seed moisture results in smaller changes in RH than in the linear region. Measurement of RH is also less accurate at the higher RH.

Near-Infrared Spectroscopy

Norris (1964) applied the principles of near-infrared reflectance spectroscopy (NIRS) to construct an apparatus to determine the moisture content of ground grain. That technique was later expanded to include the determination of oil and protein. Commercial equipment was introduced in the 1970s and several models are now widely used in the grain industry for measuring protein, moisture, and oil content (Federal Grain Inspection Service, 1987). The method is rapid, requires no chemical reagents, and is accurate when properly conducted.

In NIRS, a photodetector measures the amount of filtered light reflected by the grain sample at specific wavelengths. Protein, oil, and moisture molecules have differing reflectance characteristics at various wave bands in the near-infrared spectrum (Trevis, 1974). Absorption at 1.93μ is used to determine moisture contents of various grains (Law & Tkachuck, 1977b; Hart et al., 1962).

The NIRS equipment must be carefully calibrated against a standard reference method for each kind of seed. Of all the factors that may affect accuracy and precision, particle size is the most important (Williams, 1975). Therefore, it is necessary to use the same grinder during calibration and during sample runs. Law and Tkachuck (1977a) described a method that is independent of particle size. Williams and Cordeiro (1979) determined that frost, immaturity, diseases, and several other quality factors had little or no effect on accuracy of determination.

Moisture loss in the grinder prescribed in the official methods of the Federal Grain Inspection Service (1987) may reach 0.02 to 0.03 g H_2O g^{-1} fw when grinding seed with 0.14 g H_2O g^{-1} fw moisture (Williams & Sigurdson, 1978). However, this grinder is prescribed because it gives the best particle size for protein analysis.

Because of the serious problems with grinding, Stermer et al. (1977) studied NIRS on whole maize and sorghum. The NIRS on whole grains is promising, especially at the upper moisture ranges where grinding is difficult and electric moisture meters are inaccurate. This procedure eliminated the inaccuracy from grinding, but increased variability of readings because of the inclusion of only a few seeds in the sample. Commercial apparatus is available that will analyze samples of whole grain as small as 5 g for protein, moisture, and oil. Errors due to grinding are eliminated and the method is completely nondestructive to the seed.

The major application of NIRS to moisture testing is in combination with protein testing in the grain industry. Only in special cases would the high cost of equipment and difficulty of calibration justify its purchase for the sole purpose of measuring seed moisture.

Pulsed Nuclear Magnetic Resonance

Miller et al. (1980) demonstrated the potential for using pulsed nuclear magnetic resonance (PNMR) spectroscopy to measure moisture content of cereal grains in the range of 0.15 to 0.40 g H_2O g^{-1} fw. Calibration curves for naturally high moisture grain were different from those for grain equilibrated to high moisture, a potential source of error in testing grain when its moisture history is unknown.

While oil content affected moisture determination in the study by Miller et al. (1980), Ratkovic (1987) described how to discriminate the proton NMR signal of seed lipids from the signal of water.

The NMR procedure is rapid, nondestructive to whole grains, and accurate over a wide range of moisture contents. The PNMR signals must be correlated with moisture values determined by standard reference methods, and calibrations need to be checked periodically. The instrument is expensive, however, and the method cannot be used when the moisture content is below 0.075 g H_2O g^{-1} fw because much of that water is bound (Miller et al., 1980).

Microwave Moisture Sensors

Absorption of microwave energy by a material can be used as an index of its water content (Pande, 1975). This principle has been utilized in developing an on-line moisture measurement system in which a controlled beam of microwave energy is passed through the product and the energy lost is measured on the other side.

Chemical Methods

One chemical method involves shaking an excess of calcium carbide with a weighed quantity of seed. The calcium carbide reacts with water to form calcium hydroxide and acetylene gas. This reaction occurs in < 30 min. The loss in weight of the sample is due to the evolution of acetylene and is used as an index of seed moisture. In a variation of this method, the sample is enclosed in a container with a pressure gauge, and the moisture content is related to the acetylene gas pressure developed (Hall, 1980).

CHOICE OF MOISTURE-TESTING METHOD

No single method can adequately meet all the moisture-testing needs of seed science and industry. Many factors must be considered when choosing the most appropriate method or equipment for the purpose intended. Among these factors are (i) seed type to be tested, (ii) range of moisture content most likely to be encountered, (iii) accuracy and precision required, (iv) time available for completing the test, (v) ease of operation and technical expertise required, (vi) destructiveness to the sample, (vii) portability and power requirements, (viii) size of sample required, and (ix) cost and availability of equipment and labor.

Of the numerous methods described, the most common primary methods are air-oven drying and Karl Fischer titration, while the most popular secondary methods are electrical and rapid drying.

When the highest level of accuracy is required for research and for calibration of air-oven procedures, the Karl Fischer method is the most suitable. For these purposes, the time required to complete a test is usually not a critical factor. The Karl Fischer method is applicable for any kind of seed over the entire moisture range from near 0 to 1 g H_2O g^{-1} fw. The equipment cost is in the intermediate range and considerable expertise and training are required to produce accurate and repeatable results.

The air-oven method is a practical alternative to the Karl Fischer method for many of the same purposes when proper procedures are used. It can also be used with any kind of seed over the entire moisture range. The procedures require little technical expertise beyond the ability to operate a precision balance. On the other hand, the drying period takes many hours for whole seeds. Standard methods that have been calibrated against Karl Fischer results must be carefully followed to achieve a high degree of accuracy. Standard methods remain to be developed for many kinds of seed. Cost is in the low to intermediate range.

When the situation requires quick results, the electrical and rapid drying methods are more practical than the primary methods. Rapid tests are needed for determining when to windrow and combine, monitoring artificial drying, and receiving seed for storage. Tests with many electrical meters can be completed in a minute or two and the machines are easy to operate with little training required. A number of models are battery operated and can be taken to the field. Prices range from a few hundred dollars for farm-type meters to a few thousand dollars for trade-type meters. Electrical meters are limited to the moisture range of about 0.06 to 0.4 g H_2O g^{-1} fw, however. They are not as accurate as Karl Fischer and air-oven methods, especially above about 0.25 g H_2O g^{-1} fw. Electrical meters are best adapted for use with free-flowing grains and seeds. Capacitance meters are not adapted to moisture testing of chaffy grass seeds.

Rapid drying methods are useful when seed moisture content is beyond the range of measurement by electrical meters. They may also be used for nonfree-flowing seeds that cannot be tested with electrical meters. These are simple to operate but may require up to 30 or 40 min to complete a test. Accuracy is not as good as with the oven method. Costs are in the low to intermediate range.

Special situations exist where the more common moisture-testing methods do not serve the purpose. A nondestructive technique would be desirable in the case of gene banks where small samples are sometimes available. In such cases, NMR and NIR transmission spectroscopy are available. Hygrometric meters might offer a lower cost alternative. Continuous monitoring of moisture content on a combine or in a conditioning line is available with infrared, radio frequency, or microwave sensors.

FUTURE NEEDS

The measurement of moisture content is a well-developed science in many industries. In agriculture, methods for testing moisture content of cereals and flour are accurate, precise, and fast. Much remains to be done to develop adequate moisture-testing procedures for the seed industry. In particular, I would note the following needs of a general nature:

1. Oven-methods of moisture testing should be incorporated in the Rules for Testing Seeds of the AOSA. Such methods would serve as a guide for conducting moisture tests in research as well as in industry. All recommended methods should be proven against standard reference methods. The methods listed in Table 1 could serve as a starting place.

2. The oven methods of the ISTA should be verified by comparison with standard reference methods. Many additional species need to be added to the rules.
3. The practices of two-stage drying and grinding should be reevaluated since they are potential sources of serious error in oven-moisture test results. Methods for testing whole seed should be available as optional methods for use when time is not a limiting factor.
4. Fast and accurate methods for testing nonfree-flowing seeds need to be developed. Most electrical meters on the market were developed for cereals and other free-flowing seeds and are not adapted for use with grasses and seeds with similar characteristics.
5. Rapid moisture meters are needed that are effective over a wider moisture range than that now measured by conductance and capacitance meters.

I am sure that many needs of a more specific nature will occur to the readers of this publication.

REFERENCES

American Association of Cereal Chemists. 1979. Cereal laboratory methods. AACC Method 44-15A, moisture-air-oven methods; AACC method 44-18, moisture-modified two-stage air-oven method; AACC method 44-19, moisture-air-oven method, drying at 135°C; AACC method 44-40, moisture-modified vacuum-oven method. AACC, St. Paul.

American Society of Agricultural Engineers. 1987. ASAE standards 1987. ASAE, St. Joseph, MI.

Association of Official Analytical Chemists. 1984. Official methods of analysis. AOAC method 14.063. p. 249. AOAC, Washington, DC.

Association of Official Seed Analysts. 1981. Rules for testing seeds. J. Seed Technol. 6:1-125.

Backer, J.F., and A.W. Walz. 1985. Seed moisture testing—cook one up in your kitchen. Crops Soils 37(Dec.):15-16.

Becker, H.A., and H.R. Sallans. 1956. A study of the desorption isotherms of wheat at 25°C and 50°C. Cereal Chem. 33:79-91.

Benjamin, E., and D.F. Grabe. 1988. Development of oven and Karl Fischer techniques for moisture testing of grass seeds. J. Seed Technol. 12(1):76-89.

Bidwell, G.L., and W.F. Sterling. 1925. Preliminary notes on the direct determination of moisture. J. Assoc. Off. Agric. Chem. 8:295-301.

Bonner, F.T. 1972. Measurement of moisture content in seeds of some North American hardwoods. Proc. Int. Seed Test. Assoc. 37:975-983.

----. 1974. Determining seed moisture in *Quercus*. Seed Sci. Technol. 2:399-405.

----. 1979. Measurement of seed moisture in *Liriodendron, Prunus* and *Pinus*. Seed Sci. Technol. 7:277-282.

Buszewicz, G. 1962. A comparison of methods of moisture determination for forest seeds. Proc. Int. Seed Test. Assoc. 27:952-961.

Dexter, S.T. 1947. A method for rapidly determining the moisture content of hay or grain. Mich. Agric. Exp. Stn. Q. Bull. 30:158-166.

----. 1948. A colorimetric test for estimating percent moisture in the storage of farm products or other dry materials. Mich. Agric. Exp. Stn. Q. Bull. 30:422-426.

Federal Grain Inspection Service. 1987. Grain inspection handbook. USDA, Fed. Grain Inspection Serv., Washington, DC.

Fischer, K. 1935. A new method for analytical determination of the water content of liquids and solids. Angew. Chem. 48:394-396.

Grabe, D.F. 1984. Report of the Seed Moisture Committee 1980-1983. Seed Sci. Technol. 12:219-226.

----. 1987. Report of the Seed Moisture Committee 1983-1986. Seed Sci. Technol. 15:451-462.

Guilbot, A., J.L. Multon, and G. Martin. 1973. Determination de al teneur en eau des semences. Seed Sci. Technol. 1:587-611.

Hall, C.W. 1980. Drying and storage of agricultural crops. AVI Publ. Co., Westport, CT.

Hart, J.R. 1972. Effect of loss of nonaqueous volatiles and of chemical reactions producing water on moisture determinations in corn. Cereal Sci. Today 17:10-13.

----, L. Feinstein, and C. Golumbic. 1959. Oven methods for precise measurement of moisture content of seeds. Mktg. Res. Rep. 304. USDA, Washington, DC.

----, and C. Golumbic. 1962. A comparison of basic methods for moisture determination in seeds. Proc. Int. Seed Test. Assoc. 27:907-919.

----, and ----. 1963. Methods of moisture determination in seeds. Proc. Int. Seed Test. Assoc. 28:911-933.

----, and ----. 1966. The use of electronic moisture meters for determining the moisture content of seeds. Proc. Int. Seed Test. Assoc. 31:201-212.

----, and M.H. Neustadt. 1957. Applications of the Karl Fischer method to grain moisture determination. Cereal Chem. 34:26-37.

----, K.H. Norris, and C. Golumbic. 1962. Determination of the moisture content of seeds by near-infrared spectrophotometry of their methanol extracts. Cereal Chem. 39:94-99.

Hlynka, I., and A.D. Robinson. 1954. Moisture and its measurement. p. 1-45. *In* J.A. Anderson and A.W. Alcock (ed.) Storage of cereal grains and their products. AACC, St. Paul.

Hunt, W.H. 1965. Problems associated with moisture determination in grain and related crops. p. 123-125. *In* A. Wexler (ed.) Humidity and moisture measurement and control in science and industry. Vol. 2. Reinhold, New York.

----, and M.H. Neustadt. 1966. Factors affecting the precision of moisture measurement in grains and related crops. J. Assoc. Off. Anal. Chem. 49:757-763.

----, and S.W. Pixton. 1974. Moisture, its significance, behavior, and measurement. p. 1-39. *In* C.M. Christensen (ed.) Storage of cereal grains and their products. 2nd ed. AACC, St. Paul.

Hurburgh, C.R., Jr., T.E. Hazen, and C.J. Bern. 1985. Corn moisture measurement accuracy. Trans. ASAE 28:634-640.

----, L.N. Paynter, S.G. Schmitt, and C.J. Bern. 1986. Performance of farm-type moisture meters. Trans. ASAE 29:1118-1123.

International Association for Cereal Chemistry. 1976. ICC Standard 109/1, determination of the moisture content of cereals and cereal products (Basic reference method); ICC Standard 110/1, determination of the moisture content of cereals and cereal products (Practical method). Int. Assoc. Cereal Chem., Schwechat, Austria.

International Organization for Standardization. 1985. ISO 711-1985(E), cereals and cereal products—determination of moisture content (basic reference method); ISO 712-1985(E), cereals and cereal products—determination of moisture content (routine reference method). ISO, Geneva, Switzerland.

International Seed Testing Association. 1931. International rules for seed testing. Proc. Int. Seed Test. Assoc. 3:333.

----. 1966. International rules for seed testing. Proc. Int. Seed Test. Assoc. 31:128-134.

----. 1985. International rules for seed testing. Seed Sci. Technol. 13:338-341, 493-495.

Jones, F.E., and C.S. Brickenkamp. 1981. Automatic Karl Fischer titration of moisture in grain. J. Assoc. Anal. Chem. 64:1277-1283.

Justice, O.L., and L.N. Bass. 1978. Principles and practices of seed storage. U.S. Gov. Print. Office, Washington, DC.

Klein, L.M., and J.E. Harmond. 1971. Seed moisture—A harvest timing index for maximum yields. Trans. ASAE 14:124-126.

Kostryko, K., and T. Plebanski. 1965. Improved apparatus for moisture extraction from friable materials. p. 27-34. *In* A. Wexler (ed.) Humidity and moisture measurement and control in science and industry. Vol. 4. Reinhold. New York.

Law, D.P., and R. Tkachuk. 1977a. Determination of moisture content in wheat by near-infrared diffuse reflectance spectrophotometry. Cereal Chem. 54:874-881.

----, and ----.. 1977b. Near infrared diffuse reflectance spectra of wheat and wheat components. Cereal Chem. 54:256-265.

Miller, B.S., M.S. Lee, J.W. Hughes, and Y. Pomeranz. 1980. Measuring high moisture content of cereal grains by pulsed nuclear magnetic resonance. Cereal Chem. 57:126-129.

Mitchell, J., Jr., and D.M. Smith. 1977. Aquametry. Part I. John Wiley and Sons, New York.

----, and ----.. 1980. Aquametry. Part III. John Wiley and Sons, New York.

Multon, J.L. 1979. International standardized methods and moisture meters for determining moisture content in cereal grains. Cereal Foods World 24:548-558.

Nelson, S.O. 1981. Review of factors influencing the dielectric properties of cereal grains. Cereal Chem. 58:487-492.

Norris, K.H. 1964. Reports on design and development of a new moisture meter. Agric. Eng. 45:370.

Pande, A. 1974. Handbook of moisture determination and control. Vol. 1. Marcel Dekker, New York.

----. 1975. Handbook of moisture determination and control. Vol. 2. Marcel Dekker, New York.

Paynter, L.N., and C.R. Hurburg, Jr. 1983. Reference methods for corn moisture determination. ASAE Summer Meet., Montana State Univ., Bozeman. June. ASAE Paper 83-3088. ASAE, St. Joseph, MO.

Ratkovic, S. 1987. Proton NMR of maize seed water: The relationship between spin-lattice relaxation time and water content. Seed Sci. Technol. 15:147-154.

Roberts, E.H. 1972. Viability of seeds. Syracuse Univ. Press, Syracuse, NY.

Robertson, J.A., and W.R. Windham. 1983. Automatic Karl Fischer titration of moisture in sunflower seed. J. Am. Oil Chem. Soc. 60:1773-1777.

Scholz, E. 1981. Pyridine-free Karl Fischer reagents. Am. Lab. 13(8):89, 91.

Shanbhag, S., M.P. Steinberg, and A.I. Nelson. 1970. Bound water defined and determined at constant temperature by wide-line NMR. J. Food Sci. 35:612-615.

Stermer, R.A., Y. Pomeranz, and R.J. McGinty. 1977. Infrared reflectance spectroscopy for estimation of moisture in whole grain. Cereal Chem. 54:345-351.

Trevis, J.E. 1974. Seven automated instruments. Cereal Sci. Today 19:182-189.

U.S. Department of Agriculture. 1959. Methods for determining moisture content as specified in the official grain standards of the United States and in the United States standards for beans, peas, lentils, and rice. USDA Serv. and Reg. Ann. 147. U.S. Gov. Print. Office, Washington, DC.

Vertucci, C.W., and A.C. Leopold. 1984. Bound water in soybean seed and its relation to respiration and imbibitional damage. Plant Physiol. 75:114-117.

Wiese, E.L., R.W. Burke, and J.K. Taylor. 1965. Gas chromatographic determination of the moisture content of grain. p. 3-6. *In* A. Wexler (ed.) Humidity and moisture measurement and control in science and industry. Vol. 4. Reinhold, New York.

Williams, P.C. 1975. Application of near infrared reflectance spectroscopy to analysis of cereal grains and oilseeds. Cereal Chem. 52:561-576.

----, and H.M. Cordeiro. 1979. Determination of protein and moisture in hard red spring wheat by near-infrared reflectance spectroscopy. Influence of degrading factors, dockage, and wheat variety. Cereal Foods World 26:124-128.

----, and J.T. Sigurdson. 1978. Implications of moisture loss in grains incurred during sample preparation. Cereal Chem. 55:214-229.

Winston, P.W., and D.H. Bates. 1960. Saturated solutions for the control of humidity in biological research. Ecology 41:232-237.

Zeleny, L. 1953. Determination of moisture content of seeds. Proc. Int. Seed Test. Assoc. 18:130-141.

6 The Kinetics of Seed Imbibition: Controlling Factors and Relevance to Seedling Vigor

Christina W. Vertucci

USDA-ARS
National Seed Storage Laboratory
Fort Collins, Colorado

The early stages of seed hydration, a process known as *imbibition,* mark the period when a seed changes from an anhydrous to a fully hydrated organism capable of growing and responding to environmental stimuli. It is believed that this transition from anhydrobiosis to life in the hydrated state involves considerable reorganization of cellular constituents (Larson, 1968; Simon, 1974; Simon & Raja Harun, 1972; Leopold, 1980).

As the seed hydrates, it becomes particularly sensitive to cool temperatures and rapid imbibition and may leak solutes and macromolecules profusely, or may reinitiate faulty metabolism (Vertucci & Leopold, 1986; Vertucci & Leopold, 1984; Duke & Kakefuda, 1981; Leopold, 1980; Powell & Matthews, 1978; Bramlage et al., 1978; Simon & Raja Harun, 1972; Hobbs & Obendorf, 1972). Based primarily on leakage studies, it is surmised that the stresses of imbibition interfere with the re-establishment of cellular organelles—particularly the membranes (Simon, 1974; Simon & Raja Harun, 1972; Larson, 1968; Leopold, 1980; Vertucci & Leopold, 1984; Vertucci & Leopold, 1986).

The sensitivity of seeds to imbibitional stress is controlled by three factors: the initial moisture content of the seed, the temperature of the medium, and the rate at which water is taken up (Pollock, 1969). While the first two components are controlled by the environment, the last factor is a complicated function of both the environment and the intrinsic properties of the seed. The interaction of initial moisture content, temperature, and rate of imbibition has dramatic effects on subsequent seedling vigor (Fig. 6–1). When seeds with low initial moisture contents are imbibed rapidly at 25 °C, there is a slight decline in vigor. Reductions in vigor are exacerbated if imbibition occurs quickly at 5 °C. However, if imbibed slowly, there are no deleterious effects of imbibition at low temperatures.

The factors governing the rate at which seeds hydrate have remained elusive, in spite of many studies. This chapter reviews the kinetics of imbibi-

Copyright © 1989 Crop Science Society of America, 677 S. Segoe Rd., Madison, WI 53711, USA. *Seed Moisture,* CSSA Special Publication no. 14.

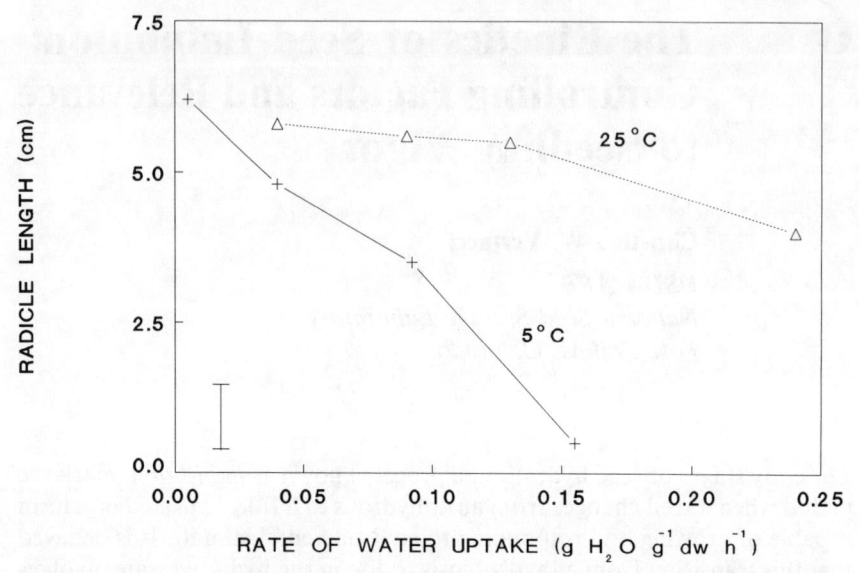

Fig. 6-1. The effect of imbibition rate at 5 and 25 °C on soybean (cv. Wayne) seed vigor. Seeds were imbibed in petri plates at indicated temperatures until moistures reached 1 g H_2O g^{-1} dw, at which point they were rolled in paper towels and incubated at 25 °C. Imbibition rates were controlled by adding various quantities of water and paper towels to the petri plates. Vigor is expressed as radicle length after 72 h incubation at 25 °C. Error bars represent the maximum standard error of the mean of 50 seeds.

tion in order to understand the parameters controlling water uptake rates and how these are related to the success of subsequent germination.

IMBIBITION KINETICS: AN OVERVIEW

Water uptake during seed germination is generally classified into three phases: rapid hydration, a lag period, and another phase of rapid hydration marked by radical emergence (Fig. 6-2). Imbibition is identified with the first stage of water uptake, and is regarded as a physical process: the consequence of matric forces of the seed and the water availability.

As a physical process, imbibition can, and has been, described by the laws of diffusion and hydraulic flow. Both treatments are valid, differing mainly in how a water gradient is expressed: in terms of concentration (diffusion) or water potential (hydraulic flow). Water moves from a high concentration or high water potential to a low concentration or low water potential. Thus, water uptake by seeds can be described by Fick's laws of diffusion (and the continuity equation):

$$J = -D \, (dc/dx) \qquad [1]$$

$$dJ/dx = -dc/dt \qquad [2]$$

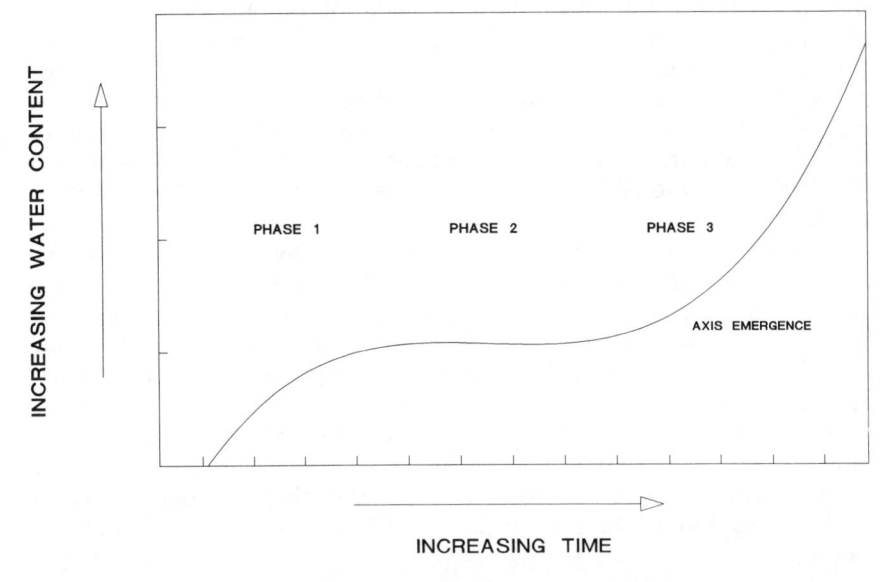

Fig. 6-2. Characteristic curve of water uptake vs. germination time for seeds describing three phases of water uptake: (1) an initial phase of rapid water absorption, (2) a lag phase, and (3) a second period of rapid water uptake marked by radicle emergence. Imbibition period is described by phase 1.

$$dc/dt = D\,(d^2c/dx^2) \qquad [3]$$

where J is the flux in grams of water cm^{-2} s^{-1}, D the diffusion coefficient in cm^2 s^{-1}, and dc/dx the water concentration gradient. Fick's first law (Eq. [1]) states that water moves down a concentration gradient, and that the rate of water flux is proportional to the concentration gradient and a factor related to the permeability of the seed. The continuity equation (Eq. [2]) states that, as the concentration of the water increases with time, the flux decreases as water penetrates the seed. When Fick's first law is substituted into the continuity equation, we obtain Fick's second law (Eq. [3]) for the special case where D is constant.

Imbibition by seeds has also been described by Darcy's law of hydraulic flow through a porous medium:

$$Q = K\,(\Delta\psi_{m-s}) \qquad [4]$$

where Q is the flow rate in cm^3 (cm^2-s)$^{-1}$ or cm s^{-1}, K is the hydraulic conductivity coefficient in cm s^{-1} MPa^{-1}, and $\Delta\psi_{m-s}$ is the water potential difference between the seed (ψ_s) and its surroundings (ψ_m). In this case, the water gradient is expressed in energy terms—water potential. To determine the water-potential gradient, we must consider the total water potential of both the soil and seed system. Total water potential of the seed (ψ_s) is composed of several terms, including osmotic, matric, and turgor pressures. In dry seeds, matric and osmotic potentials primarily contribute to the total water potential. The matric potential will be a function of the water-binding capacity of macromolecules. Proteins and insoluble carbohydrates, which are hydrophilic, are primarily responsible for the absorption of water; lipid moieties, which are mostly hydrophobic, will sorb little water (Smith & Circle, 1972; Vertucci & Leopold, 1987). The osmotic potential will be a function of the concentration of low-molecular-weight solutes. In soils, matric potential is likely to be most important in determining total water potential of the medium (ψ_m).

An exponential relationship exists between water potential and water content and is describable by seed moisture isotherms as shown in Fig. 6-3 (Vertucci & Leopold, 1987). Thus, by applying various correction factors, Eq. [1] can be used to derive Eq. [4]. Consequently, one arrives at similar conclusions by approaching imbibition from the perspective of either a diffusion process or a hydraulic process: either a water concentration gradient or a water-potential gradient serves as the driving force for imbibition and either the diffusion coefficient or the hydraulic conductivity coefficient (the term *diffusivity* will be used to refer to either of these coefficients) serves as a permeability factor that further modifies the rate at which water is absorbed.

IMBIBITION KINETICS: EFFECT OF THE MOISTURE GRADIENT

Both Fick's first law of diffusion (Eq. [1]) and Darcy's law for unsaturated water flow (Eq. [4]) state that the rate of water uptake is controlled

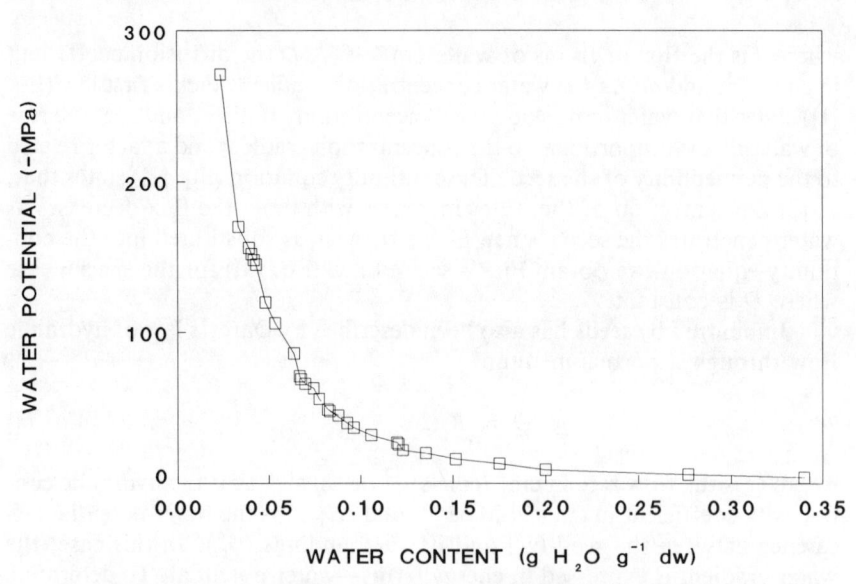

Fig. 6-3. The relationship between water content and water potential for soybean seeds. Seeds were equilibrated at 25 °C over saturated solutions that controlled relative humidity. Water potential was then determined by the equation $\psi_s = -RT/V [\ln(aw)]$ where R is the ideal gas constant, T is the absolute temperature, V is the molar volume of water, and aw is the water activity (RH/100). Data are taken from Vertucci and Leopold (1987).

by a gradient. In fact, the gradient is the driving force for the reaction. Both Eq. [1] and [4] imply that when water is in ample supply, it will be absorbed by the dry seed, provided the seed is permeable. The continuity equation predicts a rapid phase of water uptake followed by a much slower phase, the change of rate being due to a change in the concentration gradient with time.

Two components comprise the moisture gradient: the water concentration or potential exterior to the seed (i.e., the imbibition medium, ψ_m) and the water content or potential of the seed itself (ψ_s). Altering either of these factors is likely to change, in some predictable manner, the rate of water uptake into seeds.

The effect of the water potential of the external medium on the rate of imbibition has been studied using soil systems (Collis-George & Hector, 1966; Hadas & Russo, 1974; Dasberg, 1971); sintered glass plates under tension (Harper & Benton, 1966); solutions with various amounts of osmotica (Crocker, 1906; Uhvits, 1946; Hadas, 1976; Powell & Matthews, 1978; Woodstock and Tao, 1981); and vapor systems of differing relative humidities (RHs) (Collis-George & Melville, 1978). In all cases, decreasing water potentials (making ψ_m more negative) also decrease the rate of water uptake, presumably because $\Delta\psi_{m-s}$ becomes smaller. Unfortunately, these studies do not show if the rate reduction is proportional to the reduction in $\Delta\psi_{m-s}$. It is suggested that in most of these studies, lowering the external water potential reduces the rate of imbibition more than what would be predicted by Eq. [4]. This is because the treatments also alter components of the hydraulic conductivity coefficient. Thus, reducing soil moisture, and thereby reducing soil water potential, might also reduce the soil diffusivity and the seed-soil contact (Dasberg, 1971; Bewley & Black, 1978). Reducing the exterior water potential using osmotica such as polyethylene glycol (PEG) will drastically increase the viscosity of the imbibing medium (discussed in next section); this will also decrease diffusivity.

The water potential gradient can also be manipulated by altering the water potential of the seed. This is most easily accomplished by changing the seed moisture content, and is basically what happens as imbibition proceeds: the moisture gradient declines with a concomitant decline in water uptake rates (Fig. 6-2). However, water uptake rates are not affected in a predictable manner when the initial moisture content of the seed is manipulated. Seeds with high initial moisture imbibe faster than seeds with low-initial moisture (Fig. 6-4) even though the water potential gradient is enormous for the low-moisture seeds (about 100 MPa) and declines exponentially as the seed is wetted (Fig. 6-3) (Vertucci & Leopold, 1983; Hsu, 1983; Smith & Nash, 1961; Crean & Haisman, 1963). It has been suggested that by lowering the seed moisture content, we have decreased the seed water potential, but we have also changed the seed permeability (Vertucci & Leopold, 1983; Hsu, 1983).

It may also be possible to manipulate seed water potential by altering the chemical composition—thereby altering either the matric or osmotic potential—of the seeds. (Naturally, this argument does not apply if seeds

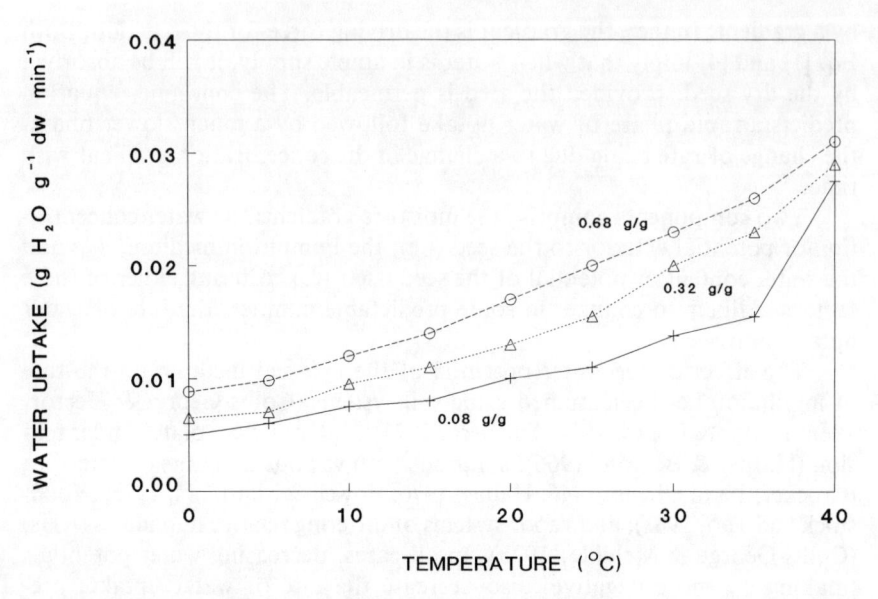

Fig. 6-4. Rate of water uptake into soybean embryos as a function of temperature and initial moisture content. Data from Vertucci and Leopold (1983).

of different chemical compositions have been equilibrated at the same RH, in which case they would have identical water potentials). For example, at any given RH, pea (*Pisum sativum* L.) seeds absorb more water than soybean (*Glycine max* L. Merr.) seeds (Vertucci & Leopold, 1987) probably because of the high starch content of pea (very negative matric potential) and the high lipid content of soybean (which probably contributes nothing to the water potential). Thus, if pea and soybean have the same initial moisture content, one would expect that pea seeds would imbibe at a faster rate than soybean seeds. This is not the case. Even with seed coats removed, it can be shown that pea seeds absorb water more slowly than soybean seeds (Tully et al., 1981). Similarly, starch from maize (*Zea mays* L.) seed endosperm is far more hydrophilic than the embryo as can be demonstrated by greater water contents at a given RH (Fig. 6-5A) (Chung & Pfost, 1967) and yet the embryo absorbs water about 10 times faster than the endosperm (Fig. 6-5B) (Stiles, 1948).

There appears to be little relationship between the water potential of seeds or seed parts and the rate at which the tissue imbibes. This suggests that the permeability of the tissue to water is quite important in determining imbibition rates.

EFFECTS OF ADDED OSMOTICA ON GERMINATION

When imbibition rates are altered by added osmotica, seed germination can be dramatically affected. Although rarely explained on the basis of im-

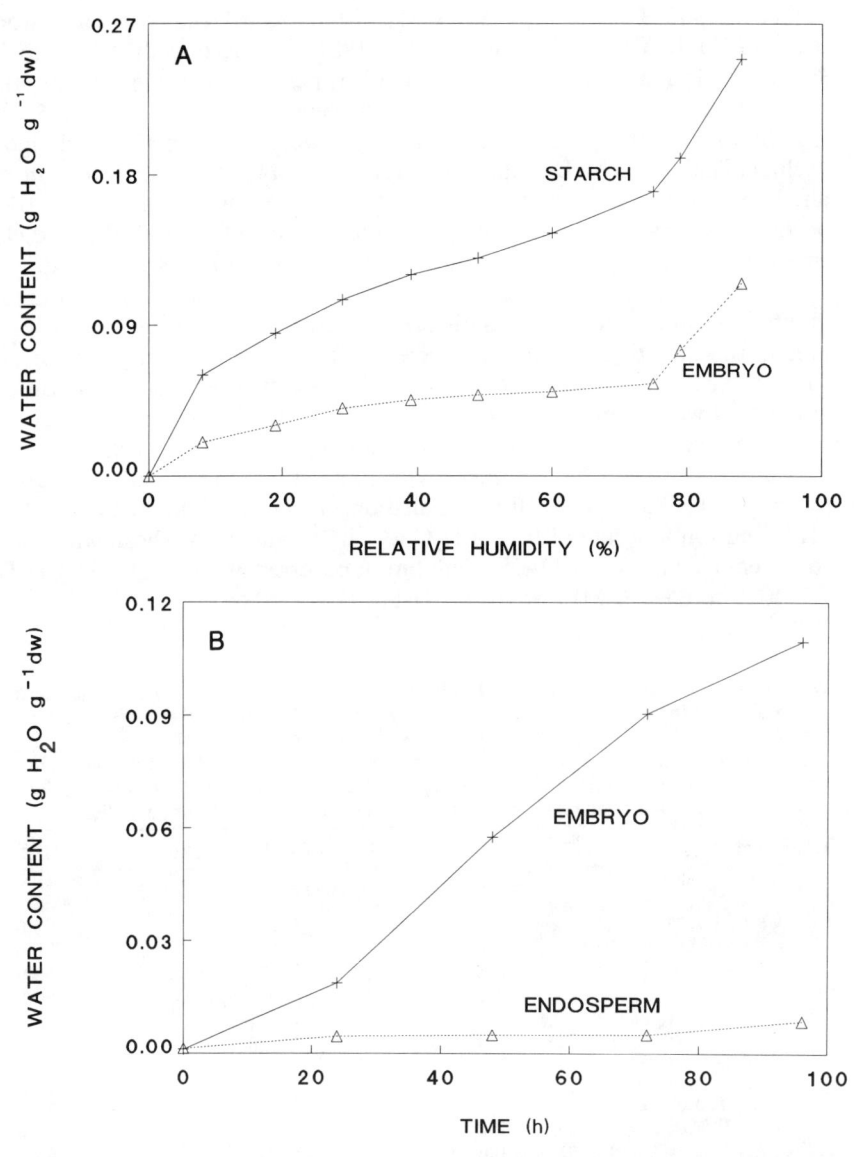

Fig. 6-5. Water sorption characteristics and rate of water uptake in the embryo and endosperm of yellow dent maize. (A) Describes the sorption isotherms of the different grain parts at room temperature. Data adapted from Chung & Pfost (1967). (B) Describes the time course of water penetration into the different grain parts. Data adapted from Stiles (1948).

bibition rates, it seems likely that the improved germination reported when seeds are imbibed in media with slightly reduced water potentials (Slosson, 1899; Uhvits, 1946; El-Sharkawi & Springuel, 1977) may be due to slightly slower rates of imbibition. The sensitivity of seeds to various imbibitional stresses can be diminished by adding osmotica. Seeds without seed coats

(Crocker, 1906; Tully et al., 1981); seeds with damaged seed coats (Johnston et al., 1979a, b; Manning & Van Standen, 1987); and aged seeds (Woodstock & Tao, 1981; Woodstock & Taylorson, 1981) perform better under osmotic stress situations than do intact or unaged seeds (for example, Table 6-1). If imbibitional rates are slowed too much, however, germination and vigor decline (Table 6-1; McDonald et al., 1987, unpublished data; Johnston et al., 1979a, b; Slosson, 1899; Uhvits, 1946; El-Sharkawi & Springuel, 1977) probably because water availability is reduced to such low levels that radicle emergence is not promoted and accelerated deterioration is enhanced.

The experiments described above presume that added osmotica control imbibition rates by decreasing moisture gradients. If this is true, then we can predict the quantity of osmotica necessary to obtain optimal germination with an experiment where rates of imbibition are measured as a function of initial seed water content, condition of testa, and PEG 8000 concentration in the imbibing medium. Rate measurements are compared with viability measurements (Table 6-1). Intact soybean at 0.15 g H_2O g^{-1} dry wt. (dw)[1] [($\psi_s = -20$ MPa) (Fig. 6-3)] performed optimally when imbibed in a -0.2 MPa solution at a rate of 0.17 g H_2O g^{-1} h^{-1}, whereas soybean with seed coats removed performed best if imbibition occurred at a rate of 0.16 g H_2O g^{-1} h^{-1} in a -1.0 MPa solution. Using Darcy's law,

Table 6-1. The effect of PEG 8000 solutions of different osmotic potentials on the imbibition rates and subsequent vigor of soybean (cv. Chippewa '64) seeds with intact seed coats and with seed coats removed. Seeds were equilibrated to about 0.15 g H_2O g^{-1} dw over a saturated NaCl solution prior to imbibition. Imbibition was considered complete when the seeds obtained 1 g H_2O g^{-1} dw (within 10-60 h). Upon completion of imbibition, seeds were rolled in paper towels and placed at 25 °C for an additional 72 h.

Treatment	ψ_m†	Rate of water uptake	GI‡
	MPa	g H_2O g^{-1} dw h^{-1}	
Intact seed coats	0	0.20	93.5
$\psi_s = -20$ MPa	-0.2	0.16	89.9
	-0.6	0.16	89.9
	-1.0	0.12	80.3
	-2.0	0.077	75.0
	-3.0	0.026	50.4
	-5.0	0.005	48.6
Seed coats removed	0	0.41	70.0
$\psi_s = -20$ MPa	-0.2	0.25	78.2
	-0.6	0.23	84.3
	-1.0	0.16	102.8
	-2.0	0.10	73.8
	-3.0	0.065	61.0
	-5.0	0.006	58.4

† Osmotic potentials of PEG solutions were calculated using the equations by Michel (1983).

‡ GI = germination index: (percent germination) × (radicle length). Standard error for the measurements of radicle lengths of 50 seedlings per treatment ranged from 7.8 to 3.4.

---

[1] To convert (g H_2O g^{-1} dw) to (%), multiply by 100. Thus, 0.10 g H_2O g^{-1} dw = 10%.

$$Q_t = K_t \, [(\Delta\psi_{m-s})_t] \quad [5]$$

$$Q_n = K_n \, [(\Delta\psi_{m-s})_n] \quad [6]$$

where Q_t and Q_n are the optimum rates for hydration of intact and coatless seeds, respectively, and are approximately the same. By equating Eq. [5] and [6] and assuming that Q_t/Q_n is equal to K_t/K_n at $\psi_m = 0$ MPa [$K_t/K_n = 0.20/0.41$ (Table 6-1)], we can solve for the optimum osmotic potential of the solution (ψ_{mn}) for imbibing coatless seeds by

$$\psi_{mn} = K_t/K_n \, [(\Delta\psi_{m-s}) + \psi_{sn}.] \quad [7]$$

For this experiment, the optimum ψ_{mn} is predicted to be about -10 MPa in contrast to the measured optimal value of about -1 MPa (Table 6-1). Note that under the experimental conditions, seeds imbibed at osmotic potentials of -5 MPa have reduced vigor. The most likely explanation for the differences between measured and predicted optimal water potential is that imbibition rates are unexpectedly much lower than predicted by a mere change in osmotic potential because PEG solutions at high concentrations are extremely viscous, and viscosity of the medium is an important component of the diffusivity coefficient.

Rates of imbibition can be manipulated by altering the moisture gradient between the imbibing medium and the seed. Few experiments however, could manipulate the imbibition rate in a manner predicted by either Eq. [1] or [4]. This is probably because water potential was not changed independently of diffusivity. Because imbibition rates cannot be manipulated predictably by changes in the moisture gradient, it is suggested that, even though the moisture gradient is the driving force for imbibition, the components of the diffusivity coefficient may be more important in determining the rate at which imbibition occurs.

IMBITION KINETICS: DIFFUSIVITY

The diffusion coefficient (D in Eq. [1]) or the hydraulic conductivity coefficient (K in Eq. [4]) are components of the flux equations that account for the permeability of the seed system to water flow. Both coefficients are complex functions of temperature, permeability of the imbibing medium and the seed, and seed/medium contact (Nobel, 1970; Bewley & Black, 1978). The effects of soil diffusivity and seed/soil contact have been handled at great length (Bewley & Black, 1978 for review, also Dasberg, 1971; Hadas & Russo, 1974; Collis-George & Hector, 1966). By imbibing seeds in water solutions, the contributions of the soil permeability and seed/soil contact to the diffusivity coefficient are eliminated. Unfortunately, in these cases, diffusivity has usually been treated as a constant.

An extension of Darcy's law, known as Poiseuille's law, demonstrates that hydraulic conductivity is actually a composite of functions of seed permeability, pore geometry, and medium viscosity:

$$Q = K' A (\Delta\psi_{m-s})/(l\, \eta) \qquad [8]$$

where K' is a coefficient related to seed permeability, A and l terms describing the dimensions of the pore, and η the viscosity of the imbibing medium.

The diffusivity of various seed species as a function of imbibition time and water content has been measured (Chittenden & Hustrulid, 1966; Shaykewich & Williams, 1971; Phillips, 1968; Sen et al., 1982; Hsu, 1983) by applying a solution to Fick's second law of diffusion (Eq. [3]) developed by Crank (1956) for spherical bodies:

$$dC/dt = 1/r^2 \; [d(r^2 D dC/dr)]/dr \qquad [9]$$

This partial second derivative can also have many solutions. One interpretation, given by Crank (1956), exists in the form

$$M_1/M_2 = \frac{\pi^2/6 \left(\sum_{n=1}^{\infty} 1/n^2 \right) \exp(-\pi^2 n^2 D t_1 / r^2)}{\pi^2/6 \left(\sum_{n=1}^{\infty} 1/n^2 \right) \exp(-\pi^2 n^2 D t_2 / r^2)} \qquad [10]$$

where M_1 and M_2 are the relative moisture contents at times t_1 and t_2, respectively, r is the radial distance from the center of the seed, and n is the number of imaginary shells of water penetration. The original derivation compares M_1/M_∞ (t_∞ = saturation time), but because seeds germinate before they stop taking up water, Crank's equation was modified so that D could be measured by comparing the moisture contents of a seed at two times, where t_1 is some time during imbibition and t_2 is often interpreted as the time required for the seed to become sufficiently hydrated to germinate, t_g. This interpretation is used commonly (Sen et al., 1982; Shaykewich & Williams, 1971; Phillips, 1968; Chittenden & Hustrulid, 1966). Hsu (1983) derives a separate set of equations with dimensionless parameters.

Crank's solution to Fick's second law makes many assumptions. First, the water uptake of the wetted region must remain constant while water penetrates the seed further. Work done by Waggoner and Parlange (1976) clearly shows that the water content at the wetting front increases. Further, Eq. [9] and [10] are written for spherical bodies of a porous matrix where water flows radially. Hence, the larger the deviation from a perfect sphere or the larger the impedance to radial flow, the more erroneous calculations of diffusivity will be. Water clearly does not flow radially in the seed as demonstrated by studies of the initial penetration of water (McDonald et al., 1987, unpublished data; Smith & Circle, 1972; Stenvert & Kingswood, 1976; Jackson & Varriano-Marston, 1980; Varriano-Marston & Jackson, 1981).

The solution to Fick's second law also assumes that the radius remains constant; that is clearly not the case for a swelling seed. Finally, Crank's derivations assume that diffusivity is constant throughout the imbibition period.

Phillips (1968) and Sen et al. (1982), calculated D for different t/t_g and M/M_g ratios. These calculations yield average values of D for t as t_1 approaches t_g. Shaykewich and Williams (1971) interpreted t_1 and t_2 as times, preferably not too far apart, at which M_1 and M_2 are known and $M_2 > M_1$, hence $t_2 > t_1$. By, differentiating Eq. [10] with respect to t, they calculated D by comparing the slopes of a water absorption curve at t_1 and t_2. Hsu (1983) described diffusivity changes as an exponential function of the initial diffusivity:

$$D = D_o \exp(kC) \qquad [11]$$

where D and D_o represent the diffusivity at time $= t$ and time $= 0$, k is a rate constant (the meaning of k is not disclosed but it does appear to be species specific), and C is the moisture concentration. The calculated diffusivities for various seeds in distilled water (Table 6-2) usually ranged from 10^{-3} to 10^{-5} cm^2 h^{-1}, although diffusivities calculated from small t_1's by Shakewhich and Williams (1971) are much lower. In most cases, diffusivity is low at low moisture contents and increases to a point as moisture increases (Fig. 6-6). The extent of change of diffusivity with moisture content varies according to the way it was determined.

Because the second law of diffusion has many solutions that become increasingly complex as the boundary conditions change, scientists investigating water infiltration into soils have attempted to provide simpler equations to describe water uptake. By assuming that there may be obstructions to water penetration and also that there is a continuous water supply, Philip (1957) developed equations, using Fick's laws and an iterative procedure that produces a simple progression series, to describe water infiltra-

Table 6-2. Calculated diffusivities of various crop species imbibed in water at room temperature.

Crop	Diffusivity	Reference
	cm^2 h^{-1}	
Soybean	10–30 × 10^{-4}	Phillips, 1968
	9–4 × 10^{-4}	Sen et al., 1982
	30 × 10^{-4}†	Hsu, 1983
Maize	5–8 × 10^{-4}	Phillips, 1968
	0.1–0.6 × 10^{-4}	Chittenden & Hustrulid, 1966
Cotton	0.2–2 × 10^{-4}	Phillips, 1968
Cowpea	4–9 × 10^{-4}	Sen et al., 1982
Rape	10^{-12} to 10^{-4}	Shaykewhich & Williams, 1971
Fava bean	30–40 × 10^{-4}	Hsu, 1983
Wheat	≈ 10^{-6}‡	Becker, 1960

† Initial moisture content was 0.55 g H$_2$O g^{-1} dw.
‡ Assumes a constant value throughout imbibition.

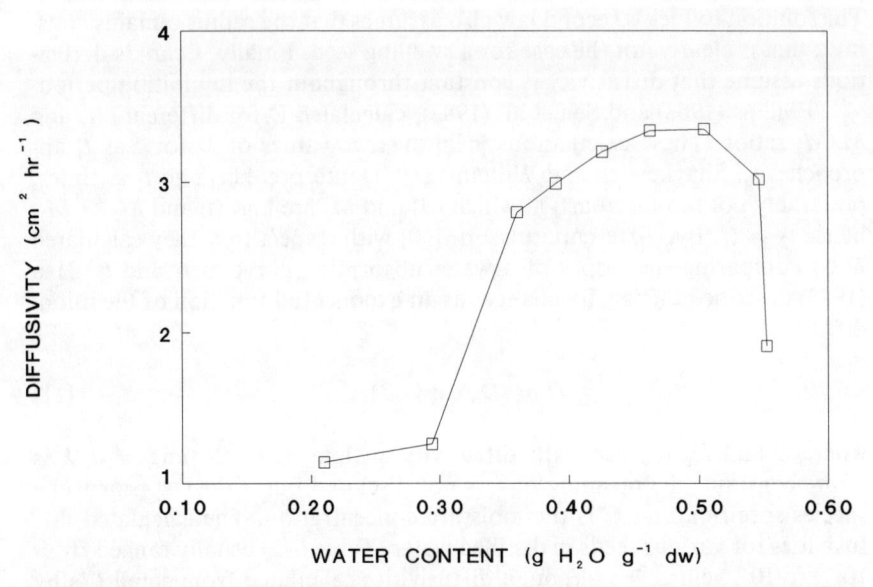

Fig. 6-6. The change in diffusivity as a function of moisture content in soybean seeds. Data derived from Phillips (1968) and assumes that M_g was equal to 0.60 g H_2O g^{-1} dw (McDonald et al., 1987, unpublished data).

tion into soils. The boundary conditions for experiments studying water infiltration into soils are similar to those for seed imbibition and so the equations developed by Philip are applicable to the study of water uptake in seeds. These equations have been simplified to

$$i = St^{1/2} + At \qquad [12]$$

$$di/dt = \tfrac{1}{2}\,(St^{1/2} + A) \qquad [13]$$

where i is water content, t is time, and S and A are constants representing absorbancy or permeability and hydraulic conductivity, respectively. Because these equations assume an infinite wetting surface, the equations only apply to seeds if we consider small times. The first term of Eq. [12] describes small times, and the second term becomes more important as t becomes greater. Data from Becker (1960) show a linear relationship between moisture gain at small times and the square root of time. If S and A values given for soils (Philip, 1957) are substituted into Eq. [12], a curve similar to the water uptake curve during imbibition of seeds results (compare Fig. 6-7 with Fig. 6-2).

The interesting point of Philip's equations is that they describe a system where there are two different coefficients for diffusivity: sorptivity and hydraulic conductivity. The factors affecting hydraulic conductivity are described in Eq. [8]. Philip (1957) describes the sorptivity coefficient by

$$S = \$ \times (\sigma \cos\theta)/\eta \qquad [14]$$

KINETICS OF SEED IMBIBITION

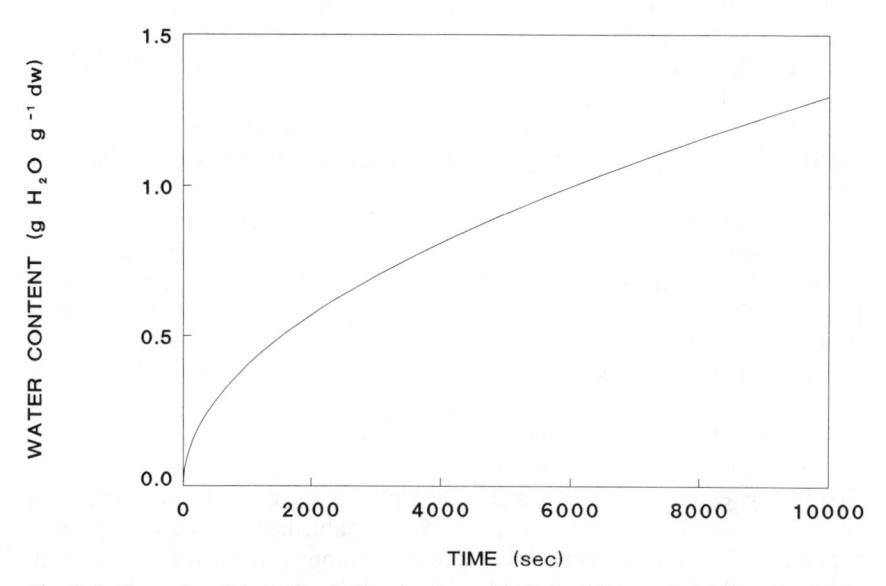

Fig. 6-7. Curve describing water infiltration into soils derived from sorptivity and hydraulic conductivity values provided by Philip (1957) and calculated using Eq. [12].

where $ is the intrinsic sorptivity (a function of geometry), σ is the surface tension of water, θ is the contact angle, and η is the viscosity. Thus, the sorptivity of a seed is a function of the seed structure and wetting ability. Equation [12] states that water infiltration can be separated into two processes: an initial wetting process and a subsequent bulk flow through the tissue.

This discussion shows that even though the diffusivity of a seed has often been treated as a constant value, it is not. Diffusivity is a function of seed moisture content, structure, and wettability as well as the medium surface tension and viscosity. Most of the seed variables are expected to change with each seed, with each seed part, or with imbibition time for a particular seed part. In particular, the diffusivity of the seed will change as imbibition progresses from the wetting stage to the hydraulic flow steps.

CHANGES IN DIFFUSIVITY WITH SEED PROPERTIES AND IMPACT ON GERMINATION

The seed coat has often been regarded as the only, or at least the primary, barrier to water. Many studies have documented that the condition of the seed coat controls the rate at which water is taken up by the seed and the success of subsequent germination (Duke et al., 1983; Duke et al., 1986; Duke & Kakefuda, 1981; Tully et al., 1981; Powell & Matthews, 1978; Arechavaleta-Medina & Snyder, 1981; Miklas et al., 1987; Taylor & Dickson, 1987; Crocker 1906; McDonald et al., 1987, unpublished data). With seed coats removed or damaged, cotyledons imbibe water rapidly and embryos are susceptible to imbibitional damage (Duke et al., 1983; Duke et al., 1986; Duke & Kakefuda, 1981; Tully et al., 1981; Powell & Matthews, 1978). Under conditions where seeds are sensitive to imbibitional damage, the diffusivity of the seed coat can be reduced by applying seed coatings. This has been accomplished by applying a thin layer of lanolin to seeds (Priestley & Leopold, 1986). It is believed that the hydrophobic lanolin reduced the wettability of the seed.

It is generally believed that thicker seed coats offer more resistance to water penetration, although there are only weak negative correlations between seed coat/embryo weight and rate of imbibition (Mugnisjah et al., 1987; Calero et al., 1981; Yaklich et al., 1986). The region around the hilum, often thought to be the thinnest (i.e., greatest diffusivity) is usually where water initially penetrates the seed (Smith & Circle, 1972; Stenvert & Kingswood, 1976; Jackson & Varriano-Marston, 1980; Varriano-Marston & Jackson, 1981). Recent evidence, however, presented by McDonald et al. (1987, unpublished data) suggested that water penetrates the region distal to the hilum first in soybean, but that this is also the thinnest region of the soybean seed coat. The work by McDonald et al. (1987, unpublished data) also suggested that the seed coat may *increase* the diffusivity (more specifically the hydraulic conductivity) of the seed by providing channels for water flow. Evidence provided by Mugnisjah (1987) suggests that although there is no correlation between seed-coat thickness and retardation of initial imbibition rates, there is a direct correlation between seed coat/embryo weight and final water uptake levels.

The morphology of the testa is likely to be critical in determining its permeability properties. For example, seeds with hard seed coats have small elongated pores and a high density of waxy material embedded in the testa epidermis (Calero et al., 1981; Yaklich et al., 1986; Agbo et al., 1987). It is suggested that the epidermis is the primary barrier to water penetration (Duke et al., 1986). Hard seeds can be made permeable by scratching or removing the seed coats (Arechavaleta-Medina & Snyder, 1981; Crocker, 1906). Dark-colored seed coats are more impermeable to water than light-colored seed coats (Tully et al., 1981; Marbach and Mayer, 1974, 1975; Mugnisjah et al., 1987; Lamprecht and Steiner, 1987), probably due to the oxidation of phenolics into hydrophobic substances (Egley et al., 1983). Oxidative reactions in the seed coat may also result in the hard-seededness (hard shell) observed when some legume seeds are stored under dry warm conditions (Vindiola et al., 1986). The incidence of seeds with seed coats of low permeability is directly related to longevity under ambient and high-temperature, high-humidity conditions (Dassou & Kueneman, 1984; Potts et al., 1978).

While permeability of seeds to water has been primarily discussed in terms of the seed coats, other parts of the seed may have differing diffusivities. The most dramatic example of this is the demonstration that different parts of the maize grain imbibed water at different rates, even though they were all equilibrated to the same water content (Stiles, 1948) (Fig. 6-5A and 6-5B). Other studies demonstrated that the axes of soybean embryos and the embryo of wheat absorbed more water more rapidly than the other grain parts (McDonald et al., 1987 unpublished data; Moss, 1977).

Although there is literature comparing diffusivities among seed species, little is known concerning the effect of seed composition of seed diffusivity. Table 6-2 illustrates that cotton (*Gossypium hirsutum* L.) has a relatively low diffusivity and soybean has a relatively high diffusivity. Since cotton is high in lipid and low in protein, it may be naturally hydrophobic. Sweet corn has a higher diffusivity than dent corn (Blacklow, 1972). The glassy nature of endosperm starch of monocots may result in small pores that block water penetration. In wheat (*Triticum aestivum* L.), protein content is negatively correlated with diffusivity (Stenvert & Kingswood, 1977; Moss, 1977; Butcher & Stenvert, 1973). However, before a generalized statement is made regarding protein content and diffusivity among species, the type of protein must be considered: it is possible that glutelins reduce diffusivity while globulins and albumins enhance it. It is suggested that seed composition is not the major determinant of diffusivity since seeds with similar compositions display remarkably different diffusivities as evidenced by differing imbibition rates (Table 6-3).

It has been suggested that the seed microstructure is critical to the seed diffusivity (Agbo et al., 1987; Stenvert & Kingswood, 1977; Moss, 1977; Butcher & Stenvert, 1973). Bean seeds with low diffusivities have small, tightly packed starch granules (Agbo et al., 1987). Wheat grains with vitreous endosperms usually hydrate slower than grains with mealy endosperms (Stenvert & Kingswood, 1977).

The idea that diffusivity of seeds depends on vigor is controversial. The hard-to-cook phenomenon, induced under accelerated aging conditions, results in the polymerization of pectic substances and gelantization of strach as well as other structural changes, but does not seem to affect the rate of water penetration (Vindiola et al., 1986; Quast et al., 1977; Moscoso et al., 1984; Varriano-Marston & Jackson, 1981; Hohlberg & Stanley, 1987). Similarly, when water uptake experiments were performed on low vigor and artificially aged (McDonald et al., 1987, unpublished data) or heat-killed (Leopold, 1980) soybean seeds, no difference in moisture uptake rates could

Table 6-3. Rates of water uptake for grain legumes with seed coats removed and initial moisture level adjusted to between 0.28 and 0.33 g H_2O g^{-1} dw. Experiments were conducted at 25 °C.

Species	Initial moisture	Rate of water uptake†
	g H_2O g^{-1} dw	g H_2O g^{-1} dw min^{-1}
Fava bean	0.28	0.949 ± 0.097
Pea	0.28	1.23 ± 0.102
Cowpea	0.31	1.44 ± 0.047
Mung bean	0.35	1.47 ± 0.103
Soybean	0.32	1.60 ± 0.041
Lima bean	0.31	2.96 ± 0.245

† Data represent mean ± SD of three measurements.

be detected when seeds were hydrated in liquid water. However, when dent corn (Blacklow, 1972), radish (*Raphanus sativus* L.), or sugar pine (*Pinus lambertiana*) embryos (Murphy & Noland 1982) were heat killed, there was an increase in the rate of water uptake. The data for rate of water absorption supplied by Blacklow (1972) are inconsistent with his calculations of the diffusivity that seemed to have been reduced by aging treatment. When accelerated and unaged soybean axes were hydrated in a humid atmosphere, there was a marked difference in the water-absorbing capabilities: the 3-d aged axes absorbed water more slowly (Fig. 6-6) (McDonald et al., 1987, unpublished data). This may be due to a change in seed diffusivity with seed deterioration.

Experiments that studied seed diffusivity as a function of imbibition time (Chittenden & Hustrulid, 1966; Phillips, 1968; Shaykewich & Williams, 1971; Sen et al., 1982) or moisture (Vertucci & Leopold, 1983; Hsu, 1983; Smith & Nash, 1961; Crean & Haisman, 1963) have demonstrated that diffusivity does change (Fig. 6-6). Although the relative change varied among experiments, diffusivity quickly increased with time when $t \ll t_g$, then gradually decreased with time as t approached t_g (Phillips, 1968; Sen et al., 1982; Shaykewich & Williams, 1971). The range in water contents for which diffusivity changes from increasing to decreasing with time appeared to be between 0.38 and 0.45 g H_2O g^{-1} for soybean (Fig. 6). This is also within the moisture content range for which the diffusivity of rape (*Brassica napus* L.) seed changed dramatically from 10^{-10} to 10^{-3} cm^2 h^{-1} (Shaykewich & Williams, 1971).

There is a difference of eight orders of magnitude between the diffusivity values for dry seeds reported by Shaykewich and Williams (1971) and those reported by others (Phillips, 1968; Sen et al., 1982; Chittenden & Hustrulid, 1966). The values presented by Shaykewich and Williams, however, are probably more representative because the computational problems experienced by other researchers were avoided by (i) calculating diffusivity using the slopes at t_1 and t_2 ($t_1 \sim t_2$) of the water absorption curve and (ii) using rape seed which does not swell (discussed later). Shaykewich and Williams (1971) implicated the seed coat as a reason for the observed abrupt increase in seed diffusivity at 0.45 g H_2O g^{-1} dw. However, if the seed coat was an important factor, an increase, not a reduction, in diffusivity would be predicted. An alternative explanation for the abrupt change in diffusivity was presented by Vertucci and Leopold (1983) who suggested that, at moistures below about 0.35 g H_2O g^{-1} dw, seed imbibition consists of a wetting process, and so the laws governing hydraulic flow do not entirely apply. This substantiates the idea that Eq. [12] and [13] are indeed applicable to seed water uptake, the first term representing a wetting effect and the second term representing bulk water flow. Thus, diffusivity is initially expressed as "sorptivity" (Eq. [14]) and later expressed as "hydraulic conductivity" (Eq. [8]). To support this argument, Vertucci and Leopold (1983) showed that water uptake rates of dry soybean cotyledons were enhanced by wetting agents, while moist cotyledons were not. Shaykewich and Williams (1971) showed that, within this initial "wetting" phase, there were large changes in the diffusivity. Hence,

KINETICS OF SEED IMBIBITION

the wetting stage is a complex process (for a further discussion see Vertucci & Leopold, 1987).

Most of the experiments described thus far were conducted at a constant temperature. When the temperature is altered, the rate at which water enters a seed is also altered—low temperatures causing slower rates (Fig. 6-4). The degree to which temperature alters the rate of imbibition has remained controversial, as has the cause of the temperature effects. Q_{10}'s >2 (Brown & Worley, 1912), suggestive of physiological or chemical processes, and Q_{10}'s of <2 (Shull, 1920; Murphy & Noland, 1982), suggestive of physical processes, have been observed in a variety of seeds. However, dramatic changes in the apparent E_a of imbibition with time (Blacklow, 1972) or initial moisture content (Vertucci & Leopold, 1983; Table 6-4) suggest that there are several mechanisms governing water uptake and each might have a different temperature coefficient. This is further demonstrated by large changes in diffusivity values with temperature: Q_{10}'s being >2 for maize (Blacklow, 1972) and apparent E_a's ranging from 12 to 4 kcal mol^{-1} for different wheat varieties (Becker, 1960).

The viscosity of water may be responsible for slowed imbibition rates at low temperatures (Allerup, 1958; Murphy & Noland, 1982; Vertucci & Leopold, 1983). This is best demonstrated during the latter stages of imbibition where hydraulic flow is important (Vertucci & Leopold, 1983; see Eq. [8]). However, during the initial wetting phase, the reciprocal of water viscosity corresponds less well to the change in water uptake, and apparent E_a's for imbibition are much higher than the E_a for viscosity changes with temperature (Vertucci & Leopold, 1983) (Table 6-4). Vertucci and Leopold (1983) suggest that temperature directly affects the wetting process. Thus, temperature may affect other elements in Eq. [14] such as the internal sorptivity ($) or the contact angle ($\theta$).

The effect of temperature on internal sorptivity ($) and its consequent effect on diffusivity has not received much attention. Although seldom considered when diffusivity calculations are made, the microstructure of seeds is important to the permeability (K' in Eq. [8] and $ in Eq. [13]) (Agbo et al., 1987; Stenvert & Kingswood, 1977; Moss, 1977; Butcher & Stenvert,

Table 6-4. Apparent E_a's calculated from Arrhenius plots of water uptake for soybean embryos of different initial moisture contents, and a comparison with temperature effects on water viscosity. Data from Vertucci & Leopold (1983).

Initial moisture	Apparent E_a
0.080	9.10
0.216	6.97
0.317	6.46
0.340	5.47
0.393	4.68
0.545	4.29
0.683	4.18
Water viscosity	−4.17

1973). This is because pore size affects the effective water flow area. Seed microstructure is likely to be altered by temperature and seed moisture content (Vertucci & Leopold, 1987; C.W. Vertucci, 1988, unpublished data).

The effect of moisture content on seed structure is manifested in how the seed swells. This has a compounding effect on calculations of seed diffusivity because it is assumed that the radius of the seed remains constant. Since most researchers (Chittenden & Hustrulid, 1966; Phillips, 1968; Sen et al., 1982; Hsu, 1983) calculated the change of diffusivity with time using Eq. [9] or [10], they used the radius of a swelled seed to calculate D when t was small, and the seed had not swelled significantly. Consequently, large errors were introduced into the calculations.

An evaluation of the error involved in calculating diffusivity using Fick's laws and a swelling material can be estimated by comparing the depth of water penetration if there was no swelling (r) and the degree of swelling a seed undergoes for every change in r (S). (Similar treatments have been conducted by Waggoner and Parlange, 1976.) Instead of describing the diffusion rate in terms of dr/dt, most calculations of diffusivity with time described the rate of water uptake in terms of dR/dt, where $R = r + S$. In this case, S is the magnitude of error in calculating r, and r is merely a function of the moisture increase with time. S is not only a function of moisture content, but it is also a function of the properties that cause swelling. If swelling is related to chemical composition, it is natural to think that different seeds will have different degrees of swelling. If $S \gg r$, and $t_1 \ll t_2$, measured diffusivities may have large errors.

To determine how the error of the calculated diffusivity changed with time, it is necessary to determine the difference between the swelling rate and the water content increase with time (i.e., compare r and R with time). This was done for soybean and radish seeds by comparing volume changes and water content changes at small times (Fig. 6-8A). It appears that during the initial stages of imbibition, the rate of relative volume increase (dV/dt) is larger than the rate of water entry (dW/dt) (Fig. 6-8B). A similar result was reported by Mugnisjah et al. (1987). As the moisture content increases to some critical value, $dV/dt - dW/dt$ also increases. Beyond a critical moisture content, $dV/dt - dW/dt$ decreases rapidly to 0. That is, the swelling occurs at the same rate as the water absorption. The critical moisture content ranges from 0.4 to 0.45 g H_2O g^{-1} dw for both radish and soybean. Again, this seems to indicate that there are at least two phases to imbibition: one where $dV/dt > dW/dt$ and one where $dW/dt = dV/dt$. Since the radius of the seed is important in the calculation of seed diffusivity and the changes in seed geometry are likely to affect diffusivity measurements, the confounding complications of seed swelling suggest that many calculations of diffusivity with water content are in error, especially when $dV/dt > dW/dt$. This is especially evident during the wetting phase of imbibition when seed-moisture contents are <0.30 g H_2O g^{-1} dw.

It is possible that the reason why $dV/dt > dW/dt$ and that $dV/dt - dW/dt$ increases with time during the wetting phase, is that biopolymers unfold allowing more sites for adsorption. Beyond a certain moisture content,

KINETICS OF SEED IMBIBITION

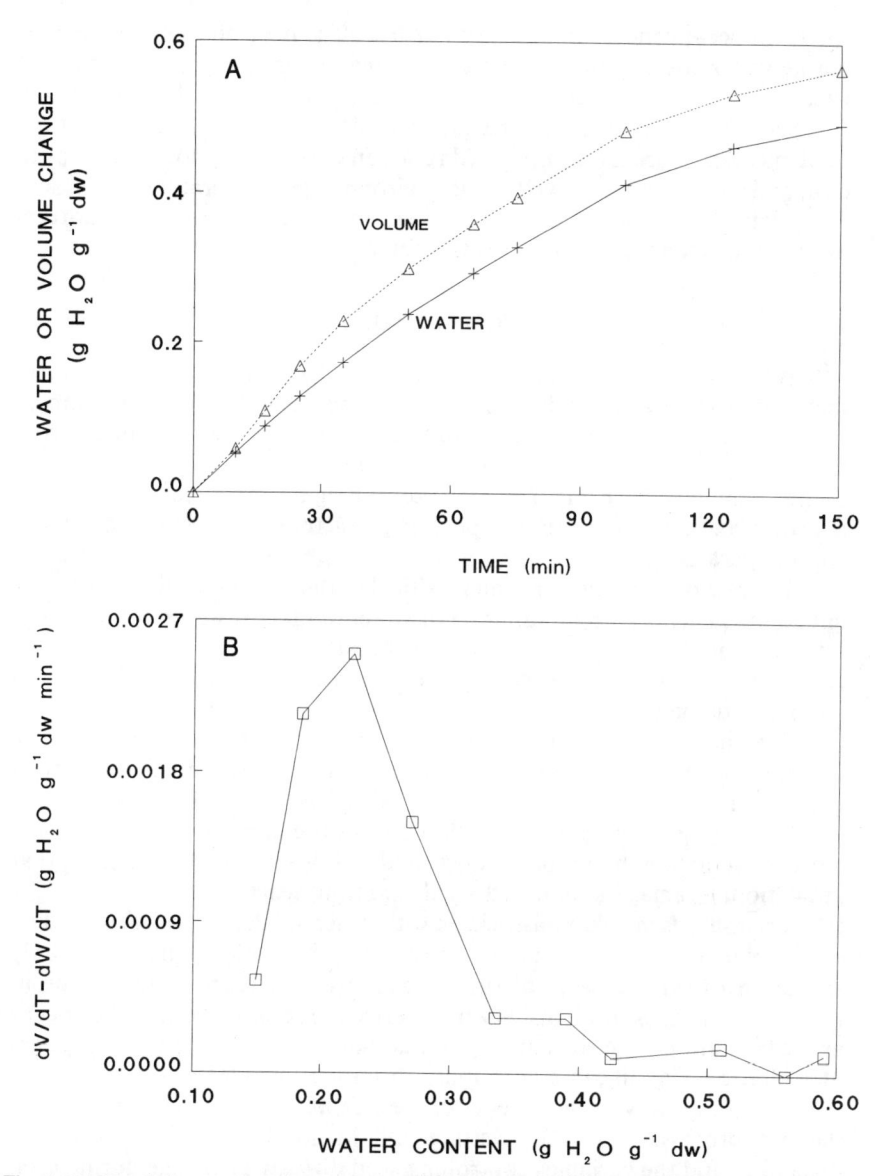

Fig. 6-8. The relationship between volume and water content changes in imbibing soybean seeds. (A) Describes the changes of water content or volume with time. Volume was measured using specific gravity flasks. Values are calculated by subtracting original water content (0.0975 g H_2O g^{-1} dw) or vol/dw (0.991 g H_2O g^{-1} dw) from subsequent measurements at the indicated time. (B) Compares the rate of change of volume with the rate of water uptake. Values are calculated by subtracting the instantaneous slope of the water uptake curve in A from the volume curve in A. The maximum standard deviation of three replicates for data given in A was ±0.001 g H_2O g^{-1} dw.

during the phase of hydraulic flow, the polymers are completely unfolded and the increase in seed volume is due primarily to the volume of water that enters the seed; hence, $dV/dt - dW/dt = 0$. The unfolding of biopolymers during water absorption seems to be affected by temperature, the extent of which also depends on seed species and tissue type (Vertucci & Leopold, 1987).

Seed diffusivity, then, is strongly dependent on the seed microstructure. Seed microstructure is a complicated function of seed morphology, composition, and water content as well as the environmental temperature. Thus, seed diffusivity is a rather complex parameter, and yet it is extremely important to our understanding of how seeds imbibe water.

CONCLUSION

We have examined the kinetics of water uptake into seeds. It is evident that imbibition is affected by the seed properties as well as the environment in which seeds germinate. A water gradient between the seed and the environment is the driving force for water uptake, but the permeability of the seed to the medium is more important in determining the rate. Seed permeability is a complex function of seed morphology, structure, composition, moisture, and temperature.

The rate of water penetration is critical to the success of subsequent germination. There is an optimum rate of water uptake. If water uptake is too slow, germinability is reduced, perhaps because of fungal infection or accelerated deterioration. If water uptake is too rapid, seeds are subject to imbibitional damage.

The changes in diffusivity with moisture content suggest that there are two phases of water uptake: an initial wetting phase and a subsequent hydraulic flow. The moisture contents at which the wetting phase is observed (<0.3 g H_2O g^{-1} dw) coincide with the moisture contents where seeds are most sensitive to imbibitional damage and so it is reasonable to suspect that imbibitional damage is imposed by the wetting reaction.

The changes in volume associated with water uptake suggest an unfolding of biopolymers as water is initially absorbed. This unfolding may reasonably account for the increases in diffusivity as moisture content increases initially. Hence, changes in diffusivity may reflect structural changes that occur as seeds hydrate. The structural changes that occur during the early stages of hydration may likewise contribute to imbibitional stresses.

The complexity of the kinetics of seed imbibition is not surprising since the two processes involved, wetting and hydraulic flow, show separate dynamics, and the variables controlling seed diffusivity change during each phase. Further, the volume of the seed undergoes changes, indicating an initial unfolding of seed polymers and then an increase proportional to the water volume. The overall kinetics of imbibition can best be understood when viewed as an integration of these two sets of variables.

REFERENCES

Agbo, G.N., G.L. Hosfield, M.A. Uebersax, and K. Klomparens. 1987. Seed microstructure and its relationship to water uptake in isogenic lines of a cultivar of dry beans (*Phaseolus vulgaris* L.). Food Microstr. 6:91-102.

Allerup, S. 1958. Effect of temperature on uptake of water in seeds. Physiol. Plant. 11:99–105.
Arechavaleta-Medina, F., and H.E. Snyder. 1981. Water imbibition by normal and hard soybean. J. Am. Oil Chem. Soc. 58:976–979.
Becker, H.A. 1960. On the absorption of liquid water by the wheat kernel. Cereal Chem. 37:309–323.
Bewley, J.D., and M. Black. 1978. Physiology and biochemistry of seeds. Springer-Verlag, New York.
Blacklow, W.M. 1972. Mathematical description of the influence of temperature and seed quality on imbibition by seeds of corn (*Zea mays* L.). Crop Sci. 12:643–646.
Bramlage, W.J., A.C. Leopold, and D.J. Parrish. 1978. Chilling stress to soybeans during imbibition. Plant Physiol. 61:525–529.
Brown, A.J., and F.P. Worley. 1912. The influence of temperature on the absorption of water by seeds of barley in relation to the temperature coefficient of chemical change. Proc. R. Soc. London B 85:546–553.
Butcher, J., and N.L. Stenvert. 1973. Conditioning studies on Australian wheat. III. The role of the rate of water penetration into the wheat grain. J. Sci. Food Agric. 24:1077–1084.
Calero, E., S.H. West, and K. Hinson. 1981. Water absorption of soybean seeds and associated causal factors. Crop Sci. 21:926–933.
Chittenden, D.H., and A. Hustrulid. 1966. Determining drying constants for shelled corn. Trans. ASAE 9:52–55.
Chung, D.S., and H.B. Pfost. 1967. Adsorption and desorption of water vapor by cereal grains and their products. Trans. ASAE 10:549–555.
Collis-George, N., and J.B. Hector. 1966. Germination of seeds as influenced by matric potential and by area of contact between seed and soil water. Aust. J. Soil Res. 4:145–164.
----, and M.D. Melville. 1978. Water absorption by swelling seeds. II. Surface condensation boundary condition. Aust. J. Soil Res. 16:291–310.
Crank, J. 1956. The mathematics of diffusion. Oxford Univ. Press, Oxford, England.
Crean, D.E.C., and D.R. Haisman. 1963. A note on the slow rehydration of some dried peas. Hortic. Res. 2:121–125.
Crocker, W. 1906. Role of seed coats in delayed germination. Bot. Gaz. 42:265–291.
Dasberg, S. 1971. Soil water movement to germinating seeds. J. Exp. Bot. 22:999–1007.
Dassou, S., and E.A. Kueneman. 1984. Screening methodology for resistance to field weathering of soybean seed. Crop Sci. 24:774–779.
Duke, S.H., and G. Kakefuda. 1981. Role of the testa in preventing cellular rupture during imbibition of legume seeds. Plant Physiol. 67:449–456.
----, ----, and T.M. Harvey. 1983. Differential leakage of intracellular substances from imbibing soybean seeds. Plant Physiol. 72:919–924.
----, ----, C.A. Henson, N.L. Loeffler, and N.M. Van Hulle. 1986. Role of the testa epidermis in the leakage of intracellular substances from imbibing soybean seeds and its implications for seedling survival. Physiol. Plant. 68:625–631.
Egley, G.H., R.N. Paul, K.C. Vaughn, and S.O. Duke. 1983. Role of peroxidase in the development of water impermeable seed coats in *Sida spinosa* L. Planta 157:224–232.
El-Sharkawi, H.M., and I. Springuel. 1977. Germination of some crop plant seeds under reduced water potential. Seed Sci. Technol. 5:677–688.
Hadas, A. 1976. Water uptake and germination of Leguminous seeds under changing external water potential in osmotic solutions. J. Exp. Bot. 27:480–489.
----, and D. Russo. 1974. Water uptake by seeds as affected by water stress, capillary conductivity and seed soil contact. I. Experimental study. Agron. J. 66:643–647.
Harper, J.L., and R.A. Benton. 1966. The behaviour of seeds in soil. II. The germination of seeds on the surface of a water supplying substrate. J. Ecol. 54:151–166.
Hobbs, P.R., and R.L. Obendorf. 1972. Interaction of initial seed moisture and imbibitional temperature on germination and productivity of soybean. Crop Sci. 12:664–667.
Hohlberg, A.I., and D.W. Stanley. 1987. Hard-to-cook defect in black beans. Protein and starch considerations. J. Agric. Food Chem. 35:571–576.
Hsu, K.H. 1983. A diffusion model with a concentration dependent diffusion coefficient for describing water movement in legumes during soaking. J. Food Sci. 48:618–622, 645.
Jackson, G.M., and E. Variano-Marston. 1980. A simple autoradiographic technique for studying diffusion of water into seeds. Plant Physiol. 65:1229–1230.
Johnston, S.K., D.S. Murray, and J.C. Williams. 1979a. Germination and emergence of ballonevine (*Cardiospermum halicacabum*). Weed Sci. 27:73–76.

----, R.H. Walker, and D.S. Murray. 1979b. Germination and emergence of Hemp Sesbania (*Sesbania exaltata*). Weed Sci. 27:290-293.

Lamrecht, H., and A.M. Steiner. 1987. Feuchteaustauschgeschwindigkeiten und Sorptionsisothermen bei Saatgut von Ackerbohne. Seed Sci. Technol. 15:311-323.

Larson, L.A. 1968. The effect soaking peas with and without seed coats has on seedling growth. Plant Physiol. 43:255-259.

Leopold, A.C. 1980. Temperature effects on soybean imbibition and leakage. Plant Physiol. 65:1096-1098.

Manning, J.C., and J. Van Staden. 1987. The role of the lens in seed imbibition and seedling vigour of *Sesbani punicea* (Cav.) Benth. Ann. Bot. (London) 59:705-713.

Marbach, T., and A.M. Mayer. 1974. Permeability of seed coats to water as related to drying conditions and metabolism of phenolics. Plant Physiol. 54:817-820.

----, and ----. 1975. Changes in catechol oxidase and permeability to water in seed coats of *Pisum elatius* during seed development and maturation. Plant Physiol. 56:93-96.

Michel, B.E. 1983. Evaluation of the water potentials of solutions of polyethylene glycol 8000 both in the absence and presence of other solutes. Plant Physiol. 72:66-70.

Miklas P.N., C.E. Townsend, and S.L. Ladd. 1987. Seed coat anatomy and the scarification of *Cicer* milkvetch seed. Crop Sci. 27:766-772.

Moscoso, W., M.C. Bourne, and L.F. Hood. 1984. Relationships between the hard-to-cook phenomenon in red kidney beans and water adsorption, puncture force, pectin, phytic acid and minerals. J. Food Sci. 49:1577-1582.

Moss, R. 1977. An autoradiographic technique for the location of conditioning water in wheat at the cellular level. J. Sci. Food Agric. 28:23-33.

Mugnisjah, W.Q., I. Shimano, and S. Matsumoto. 1987. Studies of the vigour of soybean seeds. II. Varietal differences in seed coat color and swelling components of seed during moisture imbibition. J. Fac. Agric. Kyushu Univ. 31:227-234.

Murphy, J.B., and T.L. Noland. 1982. Temperature effects on seed imbibition and leakage mediated by viscosity and membranes. Plant Physiol. 69:428-431.

Nobel, P.S. 1970. Introduction to biophysical plant physiology. W.H. Freeman and Co., Publ., New York.

Philip, J.R. 1957. The theory of infiltration: 4. Sorptivity and algebraic infiltration equations. Soil Sci. 84:257-264.

Phillips, R.E. 1968. Water diffusivity of germinating soybean, corn and cotton seed. Agron. J. 60:568-570.

Pollock, B.M. 1969. Imbibition temperature sensitivity of lima beans controlled by initial seed moisture. Plant Physiol. 44:907-911.

Potts, H.C., J. Duangpatra, W.G. Hairston, and J.C. Delouche. 1978. Some influences of hard-seededness on soybean seed quality. Crop Sci. 18:221-224.

Powell, A.A., and S. Matthews. 1978. The damaging effect of water on dry pea embryos during imbibition. J. Exp. Bot. 29:1215-1229.

Priestley, D.A., and A.C. Leopold. 1986. Alleviation of imbibitional chilling injury by use of lanolin. Crop Sci. 26:1252-1254.

Quast, D.G., and S.C. da Silva. 1977. Temperature dependence of hydration rate and effect of hydration on the cooking rate of dry legumes. J. Food Sci. 42:1299-1304.

Sen, H.S., A.V.S. Rao, K.R. Mahata, and P.C. Das. 1982. Water diffusivity of germinating soya-bean, cowpea and blackgram seeds at both large and small times. J. Agric. Sci. 98:209-213.

Shaykewich, C.F., and J. Williams. 1971. Resistance to water absorption in germinating rapeseed (*Brassica napus* L.) J. Exp. Bot. 22:19-24.

Shull, C.A. 1920. Temperature and rate of moisture intake in seeds. Bot. Gaz. 69:361-390.

Simon, E.W. 1974. Phospholipids and plant membrane permeability. New Phytol. 73:377-420.

----, and R.M. Raja Harun. 1972. Leakage during seed imbibition. J. Exp. Bot. 23:1076-1085.

Slosson, E.E. 1899. Alkali studies. IV. p. 5-29. *In* 9th Annu. Rep. of Wyoming Agric. Exp. Stn.

Smith, A.K., and S.J. Circle. 1972. Chemical composition of the seed. p. 61-92. *In* A.K. Smith and S.J. Circle (ed.) Soybeans: Chemistry and technology. AVI Publ. Co., Westport, CT.

----, and A.M. Nash. 1961. Water absorption of soybeans. J. Am. Oil Chem. Soc. 38:120-123.

Stenvert, N.L., and K. Kingswood. 1976. An autoradiographic demonstration of the penetration of water into wheat during tempering. Cereal Chem. 53:141-149.

----, and ----. 1977. Factors influencing the rate of moisture penetration into wheat during tempering. Cereal Chem. 54:627-637.

Stiles, I.E. 1948. Relation of water to the germination of corn and cotton seeds. Plant Physiol. 23:201–222.

Taylor, A.G., and M.H. Dickson. 1987. Seed coat permeability in semi-hard snap bean seeds: Its influence on imbibitional chilling injury. J. Hortic. Sci. 62:183–189.

Tully, R.E., M.E. Musgrave, and A.C. Leopold. 1981. The seed coat as a control of imbibitional chilling injury. Crop Sci. 21:312–317.

Uhvits, R. 1946. Effect of osmotic pressure on water absorption and germination of alfalfa seeds. Am. J. Bot. 33:278–283.

Varriano-Marston, E., and G.M. Jackson. 1981. Hard-to-cook phenomenon in beans: Structural changes during storage and imbibition. J. Food Sci. 46:1379–1385.

Vertucci, C.W., and A.C. Leopold. 1983. Dynamics of imbibition in soybean embryos. Plant Physiol. 72:190–193.

----, and ----. 1984. Bound water in soybean seed and its relation to respiration and imbibitional damage. Plant Physiol. 75:114–117.

----, and ----. 1986. Physiological activities associated with hydration level in seeds. 35–49. *In* A.C. Leopold (ed.) Membranes, metabolism, and dry organisms. Comstock Publ., Co., Ithaca, NY.

----, and ----. 1987. Water binding in legume seeds. Plant Physiol. 85:224–231.

Vindiola, O.L., P.A. Seib, and R.C. Hoseney. 1986. Accelerated development of the hard-to-cook state in beans. Cereal Foods World 31:538–551.

Waggoner, P.E., and J.Y. Parlange. 1976. Water uptake and water diffusivity of seeds. Plant Physiol. 57:153–156.

Woodstock, L.W., and K.L.J. Tao. 1981. Prevention of imbibition injury in low vigor soybean embryonic axes by osmotic control of water uptake. Physiol. Plant 51:133–139.

----, and R.B. Taylorson. 1981. Soaking injury and its reversal with polyethylene glycol in relation to respiratory metabolism in high and low vigor soybean seeds. Physiol. Plant. 53:263–268.

Yaklich, R.W., E.L. Vigil, and W.P. Wergin. 1986. Pore development and seed coat permeability in soybean. Crop Sci. 26:616–624.